THE FUNDAMENTALS OF
ELECTRON DENSITY, DENSITY MATRIX
AND DENSITY FUNCTIONAL THEORY
IN ATOMS, MOLECULES AND THE SOLID STATE

Progress in Theoretical Chemistry and Physics

VOLUME 14

Honorary Editor:

W.N. Lipscomb *(Harvard University, Cambridge, MA, U.S.A.)*

Editors-in-Chief:

J. Maruani *(Laboratoire de Chimie Physique, Paris, France)*
S. Wilson *(Rutherford Appleton Laboratory, Oxfordshire, U.K.)*

Editorial Board:

H. Ågren *(Royal Institute of Technology, Stockholm, Sweden)*
D. Avnir *(Hebrew University of Jerusalem, Israel)*
J. Cioslowski *(Florida State University, Tallahassee, FL, U.S.A.)*
R. Daudel *(European Academy of Sciences, Arts and Humanities, Paris, France)*
G. Delgado-Barrio *(Instituto de Matematicas y Fisica Fundamental, Madrid, Spain)*
E.K.U. Gross *(Freie Universität, Berlin, Germany)*
W.F. van Gunsteren *(ETH-Zentrum, Zürich, Switzerland)*
K. Hirao *(University of Tokyo, Japan)*
I. Hubač *(Komensky University, Bratislava, Slovakia)*
M.P. Levy *(Tulane University, New Orleans, LA, U.S.A.)*
R. McWeeny *(Università di Pisa, Italy)*
P.G. Mezey *(University of Saskatchewan, Saskatoon, SK, Canada)*
M.A.C. Nascimento *(Instituto de Quimica, Rio de Janeiro, Brazil)*
N. Rahman *(Dipartimento di Scienze Chimiche, Trieste, Italy)*
S.D. Schwartz *(Yeshiva University, Bronx, NY, U.S.A.)*
S. Suhai *(Cancer Research Center, Heidelberg, Germany)*
O. Tapia *(University of Uppsala, Sweden)*
P.R. Taylor *(University of Warwick, Coventry, U.K.)*
R.G. Woolley *(Nottingham Trent University, Nottingham, U.K.)*

Former Editors and Editorial Board Members:

I. Prigogine (deceased)
J. Rychlewski (deceased)
Y.G. Smeyers (deceased)
G.L. Malli (resigned)

The titles published in this series are listed at the end of this volume.

The Fundamentals of
Electron Density, Density Matrix and Density Functional Theory
in Atoms, Molecules and the Solid State

Edited by

N.I. Gidopoulos

*ISIS Facility,
Rutherford Appleton Laboratory,
Chilton, Oxfordshire, England*

and

S. Wilson

*Computational Science and Engineering Department,
Atlas Centre, Rutherford Appleton Laboratory,
Chilton, Oxfordshire, England*

KLUWER ACADEMIC PUBLISHERS
DORDRECHT / BOSTON / LONDON

A C.I.P. Catalogue record for this book is available from the Library of Congress.

ISBN 1-4020-1793-6

Published by Kluwer Academic Publishers,
P.O. Box 17, 3300 AA Dordrecht, The Netherlands.

Sold and distributed in North, Central and South America
by Kluwer Academic Publishers,
101 Philip Drive, Norwell, MA 02061, U.S.A.

In all other countries, sold and distributed
by Kluwer Academic Publishers,
P.O. Box 322, 3300 AH Dordrecht, The Netherlands.

Printed on acid-free paper

All Rights Reserved
© 2003 Kluwer Academic Publishers
No part of this work may be reproduced, stored in a retrieval system, or transmitted
in any form or by any means, electronic, mechanical, photocopying, microfilming, recording
or otherwise, without written permission from the Publisher, with the exception
of any material supplied specifically for the purpose of being entered
and executed on a computer system, for exclusive use by the purchaser of the work.

Printed in the Netherlands.

Progress in Theoretical Chemistry and Physics

A series reporting advances in theoretical molecular and material sciences, including theoretical, mathematical and computational chemistry, physical chemistry and chemical physics

Aim and Scope

Science progresses by a symbiotic interaction between theory and experiment: theory is used to interpret experimental results and may suggest new experiments; experiment helps to test theoretical predictions and may lead to improved theories. Theoretical Chemistry (including Physical Chemistry and Chemical Physics) provides the conceptual and technical background and apparatus for the rationalisation of phenomena in the chemical sciences. It is, therefore, a wide ranging subject, reflecting the diversity of molecular and related species and processes arising in chemical systems. The book series *Progress in Theoretical Chemistry and Physics* aims to report advances in methods and applications in this extended domain. It will comprise monographs as well as collections of papers on particular themes, which may arise from proceedings of symposia or invited papers on specific topics as well as initiatives from authors or translations.

The basic theories of physics – classical mechanics and electromagnetism, relativity theory, quantum mechanics, statistical mechanics, quantum electrodynamics – support the theoretical apparatus which is used in molecular sciences. Quantum mechanics plays a particular role in theoretical chemistry, providing the basis for the valence theories which allow to interpret the structure of molecules and for the spectroscopic models employed in the determination of structural information from spectral patterns. Indeed, Quantum Chemistry often appears synonymous with Theoretical Chemistry: it will, therefore, constitute a major part of this book series. However, the scope of the series will also include other areas of theoretical chemistry, such as mathematical chemistry (which involves the use of algebra and topology in the analysis of molecular structures and reactions); molecular mechanics, molecular dynamics and chemical thermodynamics, which play an important role in rationalizing the geometric and electronic structures of molecular assemblies and polymers, clusters and crystals; surface, interface, solvent and solid-state effects; excited-state dynamics, reactive collisions, and chemical reactions.

Recent decades have seen the emergence of a novel approach to scientific research, based on the exploitation of fast electronic digital computers. Computation provides a method of investigation which transcends the traditional division between theory and experiment. Computer-assisted simulation and design may afford a solution to complex problems which would otherwise be intractable to theoretical analysis, and may also provide a viable alternative to difficult or costly laboratory experiments. Though stemming from Theoretical Chemistry, Computational Chemistry is a field of research

Progress in Theoretical Chemistry and Physics

in its own right, which can help to test theoretical predictions and may also suggest improved theories.

The field of theoretical molecular sciences ranges from fundamental physical questions relevant to the molecular concept, through the statics and dynamics of isolated molecules, aggregates and materials, molecular properties and interactions, and the role of molecules in the biological sciences. Therefore, it involves the physical basis for geometric and electronic structure, states of aggregation, physical and chemical transformations, thermodynamic and kinetic properties, as well as unusual properties such as extreme flexibility or strong relativistic or quantum-field effects, extreme conditions such as intense radiation fields or interaction with the continuum, and the specificity of biochemical reactions.

Theoretical chemistry has an applied branch – a part of molecular engineering, which involves the investigation of structure–property relationships aiming at the design, synthesis and application of molecules and materials endowed with specific functions, now in demand in such areas as molecular electronics, drug design or genetic engineering. Relevant properties include conductivity (normal, semi- and supra-), magnetism (ferro- or ferri-), optoelectronic effects (involving nonlinear response), photochromism and photoreactivity, radiation and thermal resistance, molecular recognition and information processing, and biological and pharmaceutical activities, as well as properties favouring self-assembling mechanisms and combination properties needed in multifunctional systems.

Progress in Theoretical Chemistry and Physics is made at different rates in these various research fields. The aim of this book series is to provide timely and in-depth coverage of selected topics and broad-ranging yet detailed analysis of contemporary theories and their applications. The series will be of primary interest to those whose research is directly concerned with the development and application of theoretical approaches in the chemical sciences. It will provide up-to-date reports on theoretical methods for the chemist, thermodynamician or spectroscopist, the atomic, molecular or cluster physicist, and the biochemist or molecular biologist who wish to employ techniques developed in theoretical, mathematical or computational chemistry in their research programmes. It is also intended to provide the graduate student with a readily accessible documentation on various branches of theoretical chemistry, physical chemistry and chemical physics.

Contents

Preface   xi

Contributing Authors   xiii

Part I   The Workshop

The Fundamental of Electron Density, Density Matrix and Density Functional Theory for Atoms, Molecules and the Solid State   3
B. T. Sutcliffe
1   Introduction   3
2   Density Matrix Theory (DMT)   4
3   Density Functional Theory (DFT)   5
4   Conclusions   7

The Programme   9

Abstracts of Talks and Posters   15

List of Participants   33

Part II   The Proceedings

The Keldysh formalism applied to time-dependent current-density-functional theory   43
Robert van Leeuwen
1   Introduction   43
2   Keldysh action   45
3   Kohn-Sham equations and linear response   49
4   TDOPM equations   52
5   Integral equation for the xc-kernel   55
6   The exchange-only kernel for the electron gas   58
7   Conclusions   60
Appendix: A   61
Appendix: B   62

Appendix: C 63
Appendix: D 65

Towards time-dependent density-functional theory for molecules in strong laser pulses 69
T. Kreibich, N.I. Gidopoulos, R. van Leeuwen, and E.K.U. Gross
1   Introduction 69
2   Formulation of DFT approach, absence of e-n correlation functional 71
3   Ansatz to describe e-n correlation 72

Pair density functional theory 79
Á. Nagy
1   Introduction 79
2   Pair density functional theory 80
3   Discussion 86

The Kummer Variety for N-particles 89
A. J. Coleman
1   Introduction 89
2   Key Definitions 90
3   Results 92
4   Next Steps? 94

Some Unsolved Problems in Density Matrix Theory and Density Functional Theory 97
Roy McWeeny
1   Preliminaries: the density functions 98
2   Density matrices to density functionals 103
3   Examples: density functions for simple systems 105
4   A group theoretical approach 108
5   Conclusion 113

The new formulation of the density functional theory, the limitation of accuracy of the Kohn-Sham potential and its expression in terms of the external potential 115
Andreas K. Theophilou
1   Introduction 115
2   The New Formulation of DFT 116
3   The KS potential as a mapping of the external potential. 126

Functional N-representability in density matrix and density functional theory 129
E.V. Ludeña, V.V. Karasiev, P. Nieto
1   Introduction 130
2   Functional N-representability 132
3   Functional N-representability through built-in pure-state N-representability conditions 134

4		Functional N-representability in DFT: Application to Hooke's atom	136
	4.1	General considerations	136
	4.2	Constructive approach based on local-scaling transformations	137
	4.3	Construction of exact functional for Hooke's atom	138
5		Conclusions	140
6		Acknowledgments	142

Density-functional theory for the Hubbard model 145
K. Capelle, N. A. Lima, M. F. Silva, and L. N. Oliveira

1		The Hubbard model and density-functional theory	146
2		Exchange-correlation energy of the Hubbard model	149
3		Applications	154
	3.1	Luttinger liquids	154
	3.2	Impurity models	156
	3.3	Spin-density waves	160
	3.4	Mott insulator	162
4		Summary and outlook	165

Demonstrating the effectiveness of a nonlocal density functional description of exchange and correlation 169
Philip P. Rushton and Stewart J. Clark

1		Introduction	169
2		WDA theory	171
3		Cosine-wave electron gas.	172
	3.1	XC-energy density $e_{xc}(\mathbf{r})$	173
	3.2	Total XC-energy	174
	3.3	XC-potentials	175
	3.4	XC-holes	175
	3.5	Pair-correlation functions	177
4		Large amplitude perturbations	177
5		Conclusions	179
6		Acknowledgements	182

Incorporating the Virial Field into the Hartree-Fock Equations 185
R.F.W. Bader

1	Introduction	186
2	Incorporation of $V(\mathbf{r})$ into the Hartree-Fock Equations	189

Hohenberg-Kohn Theorem and Constrained Search Formulation for Diagonal Spin Density Functional Theory 195
Nikitas I. Gidopoulos

1	Introduction	195
2	Spin-Density-Fuctional Theory	196
3	Diagonal-Spin-Density-Functional theory	201

Part III   The Forum

The Forum - Questions ... 209

The Forum - Discussion ... 211

A Glossary ... 219

A Selected Bibliography ... 221

Index ... 223

Preface

This volume records the proceedings of a Forum on *The Fundamentals of Electron Density, Density Matrix and Density Functional Theory in Atoms, Molecules and the Solid State* held at the Coseners' House, Abingdon-on-Thames, Oxon. over the period 31st May - 2nd June, 2002. The forum consisted of 26 oral and poster presentations followed by a discussion structure around questions and comments submitted by the participants (and others who had expressed an interest) in advance of the meeting.

Quantum mechanics provides a theoretical foundation for our understanding of the structure and properties of atoms, molecules and the solid state in terms their component particles, electrons and nuclei. (Relativistic quantum mechanics is required for molecular systems containing heavy atoms.) However, the solution of the equations of quantum mechanics yields a function, a wave function, which depends on the coordinates, both space and spin, of all of the particles in the system. This functions contains much more information than is required to yield the energy or other property.

In an after-dinner speech delivered in June, 1959, C.A. Coulson, the first Professor of Theoretical Chemistry at the University of Oxford, said

> "It has frequently been pointed out that a conventional many-electron wave function tells us more than we need to know. All the necessary information required for the energy and for calculating the properties of molecules is embodied in the first- and second-order density matrices. "

Professor A.J. Coleman of Queen's University, Kingston, Ontario and one of the distinguished participants at the Forum, has dubbed this "Coulson's challenge".

This volume provides a record of the Forum. It consists of three parts. PART I describes the Workshop which preceded the Forum. It contains the introductory lecture delivered by Professor B.T. Sutcliffe, the Programme, the abstracts of the talks delivered and posters displayed, and a list of the participants. PART II contains contributions for the proceedings. Not all contributions correspond closely to the talks given in

Abingdon, but certainly all were inspired by the forum. PART III contains a summary of the Forum discussion. We have attempted to record as much of the discussion as possible with the help of participants.

The Forum was held in the Coseners' House, a mansion built during the reign of Queen Elizabeth I on the banks of the River Thames adjacent to the site of Abingdon Abbey, which was completely destroyed when King Henry VIII ordered the dissolution of the monasteries. The Coseners' House is built of the site of the monastery kitchens. The House and its environs provided a congenial yet stimulating venue for the meeting.

The Forum was held during the Golden Jubilee celebrations for Queen Elizabeth II. Abingdon-on-Thames claims to be the oldest town in England - older than the nearby University city of Oxford. It has many historic buildings and traditions. On the occasion of Royal celebrations buns are thrown to the populace from the magnificent County Hall which stands in the heart of the town surrounded by St. Nicholas and the Abbey Gateway to one side and St. Helen's Street leading to St. Helen's Church on the other. The County Hall lies but a hundred yards or so from the Coseners' House and participants were able to join the "bun throwing" in a break between sessions.

We are grateful to the participants not only for the high standard of the talks and posters delivered at the meeting, which is, of course, reflected in this Proceedings volume, but also for the constructive attitude with which they entered the Forum discussion.

We are grateful to the newly established CCLRC [1] *Centre for Molecular Structure and Dynamics* (CMSD) for providing funding for the forum. Indeed, the forum was the first meeting to be organized under the aupices of the Centre. We are particularly grateful to Professor C.C. Wilson, convenor of CMSD, for his support and encouragement.

We are grateful to the forum secretary, Mrs. Kathryn M. Vann, for the efficient manner in which she made preparations for the meeeting.

N.I. GIDOPOULOS, S. WILSON

---

[1] Council for the Central Laboratory of the Research Councils

Contributing Authors

R.F.W. Bader, Department of Chemistry, McMaster University, Hamilton, Ontario, L8S 4M1, Canada

K. Capelle, Departamento de Química e Física Molecular, Instituto de Química de Sao Carlos, Universidade de Sao Carlos, Caixa Postal 780, Sao Carlos, 13560-970, Brazil

S.J. Clark, Department of Physics, University of Durham, Science Laboratories, South Road, Durham DH1 3LE, England

A.J. Coleman, Department of Mathematics and Statistics, Queen's University, Canada

N.I. Gidopoulos ISIS Facility, Rutherford Appleton Laboratory, Chilton, Didcot, Oxfordshire, OX11 0QX, England

E.K.U. Gross, Institut für Theoretische Physik, Freie Universität Berlin, Arnimallee 14, 14195 Berlin, Germany

V.V. Karasiev, Centro de Química Instituto Venezolano de Investigaciones Científicas, IVIC, Apartado 21827, Caracas 1020-A, Venezuela

J. Kohanoff, Queen's University Belfast, Northern Ireland.

T. Kreibich, Institut für Theoretische Physik, Freie Universität Berlin, Arnimallee 14, 14195 Berlin, Germany

E.V. Ludeña, Centro de Química Instituto Venezolano de Investigaciones Científicas, IVIC, Apartado 21827, Caracas 1020-A, Venezuela

N.A. Lima, Departamento de Física e Informática, Instituto de Física de Sao Carlos, Universidade de Sao Carlos, Caixa Postal 369, Sao Carlos, 13560-970, Brazil

R. McWeeny, Dipartimento di Chimica e Chimica Industriale, University of Pisa, Via Risorgimento 35, 56100 Pisa, Italy

Á. Nagy, Department of Theoretical Physics, University of Debrecen, H-4010 Debrecen, Hungary

P. Nieto, Centro de Química Instituto Venezolano de Investigaciones Científicas, IVIC, Apartado 21827, Caracas 1020-A, Venezuela

L.N. Oliveira, Departamento de Física e Informática, Instituto de Física de Sao Carlos, Universidade de Sao Carlos, Caixa Postal 369, Sao Carlos, 13560-970, Brazil

P.P. Rushton, Department of Physics and Department of Chemistry, University of Durham, Science Laboratories, South Road, Durham DH1 3LE, England

M.F. Silva Departamento de Física e Informática, Instituto de Física de Sao Carlos, Universidade de Sao Carlos, Caixa Postal 369, Sao Carlos, 13560-970, Brazil

B.T. Sutcliffe, Laboratoire de Chimie Physique Moléculaire, Université Libre de Bruxelles, B-1050 Bruxelles, Belgium

A.K. Theophilou, "DEMOCRITOS" National Center for Scientific Research, TT 15310 Attica, Greece

R. van Leeuwen, Theoretical Chemistry, Material Science Centre, Rijksuniversiteit Groningen, Nijenborgh 4, 9747 AG, Groningen, The Netherlands

"It has frequently been pointed out that a conventional many-electron wave function tells us more than we need to know. All the necessary information required for the energy and for calculating the properties of molecules is embodied in the first- and second-order density matrices. These may, of course, be obtained from the wave function by a process of integration. But this is aesthetically unpleasing, and so attempts have been made ... to work directly with these matrices. There is an instinctive feeling that matters such as electron correlation should show up in the two-particle density matrix. But here we are confronted by a serious lack of success. We do know the conditions that must be satisfied by the many-electron wave function ..., but we still do not know the conditions that must be satisfied by the density matrix. "

C.A. Coulson,
Present State of Molecular Structure Calculations,
Rev. Mod. Phys. **32**, No. 2, 170 (1960)

"One of the things I found most mysterious, when I first learnt quantum mechanics, was that one did not deal with the probability density, for finding an electron at the point x. Instead, one worked with a wave function, which was a complex square root of the probability density. Of course, the fact that one deals with amplitudes or wave functions, is fundamental to quantum mechanics, because it allows the possibility of interference: probabilities are necessarily real and positive, so they can only add up. But amplitudes or wave functions, can be negative or complex, so they can cancel each other. Nevertheless, it was not clear to me, why one could describe the state of an electron by a wave function, rather than just by a probability density.

At first, I thought it must be just my stupidity, because everyone else seemed to accept wave functions without question. However, I later found that Schroedinger had had the same problem as me. When he first discovered his equation, he thought it applied to the probability density. But this did not give agreement with observation. It was only subsequently, that he realized that he could get agreement if the equation governed, not the probability density, but a complex valued quantity, whose modulus squared is the probability density.

One of the aims of this lecture, is to explain to people who are as stupid as Schroedinger and myself, why it is that one can work with amplitudes or wave functions, rather than probabilties. The reason is, that ordinary flat space time is simply connected. This means a surface of constant time, divides space time into two parts, M_+, and M_-. "

S.W Hawking,
Wormholes and Non-simply Connected Manifolds,
in *Quantum Cosmology and Baby Universes*,
edited by S. Coleman, J.B. Hartle, T. Piran and S. Weinberg,
World Scientific, Singapore (1990)

"It is intriguing to think that, if there is no representability problem in the Electron Density Functional method based on the Hohenberg-Kohn theorem, then, there can hardly be any representability problem in the approach based on the reduced density matrices. Personally, I do not believe this is the case."

P.-O. Löwdin (1987)
Quoted in E.S. Kryachko and E.V. Ludeña,
Energy Density Functional Theory of Many-Electron Systems,
Kluwer Academic Publishers, 1990

"I begin with a provocative statement. *In general the many-electron wave function* $\Psi(r_1,...,r_N)$ *for a system of N electrons is not a legitimate scientific concept, when* $N \geq N_0$ *, where* $N_0 \approx 10^3$. I will use two criteria for defining "legitimacy": (a) That Ψ can be calculated with sufficient accuracy and (b) that $\Psi(r_1,...,r_N)$ can be recorded with sufficient accuracy. ... Having attempted to discredit the very-many-electron wavefunction $\Psi(r_1,...,r_N)$, for *many* electrons I must, of course, recall two well-known facts: physically/chemically interesting quantities, like total energy E, density $n(r)$, pair correlation function $g(r,r')$, etc., depend on only very *few variables* and, formally, can be thought of as obtained by tracing over all other variables, e.g.,

$$n(r) = N \int \Psi^*(r,r_2,...,r_N)\Psi(r,r_2,...,r_N) \, dr_2 \ldots dr_N \, ;$$

and that some $\tilde{\Psi}$'s which, ... are hopelessly "bad" for large N, can give respectable and even accurate results for those contracted quantities. Of course not *every* bad trial $\tilde{\Psi}$ will give good results for those quantities, and the question of how one discriminates the useful "bad" $\tilde{\Psi}$, from the vast majority of useless "bad" $\tilde{\Psi}$'s requires much further thought. "

W. Kohn,
Electronic structure of matter (Nobel Lecture),
Rev. Mod. Phys. Vol. 71, No 5, October 1999.
(Copyright The Nobel Foundation 1998)

I

THE WORKSHOP

THE FUNDAMENTALS OF ELECTRON DENSITY, DENSITY MATRIX AND DENSITY FUNCTIONAL THEORY FOR ATOMS, MOLECULES AND THE SOLID STATE - A FORUM PREVIEW

B. T. Sutcliffe
Laboratoire de Chimie Physique Moléculaire,
Université Libre de Bruxelles,
B-1050 Bruxelles, Belgium
bsutclif@ulb.ac.be

Abstract The topics to be discussed at this meeting are surveyed and an attempt is made to provide a context in which they might be placed.

1. Introduction

When the organisers invited me to provide an introduction to this symposium I at first demurred. I did so because I imagined that during the introductory session all in the audience would be wondering why I, a person largely unqualified to do so, was performing. I remembered the scornful intervention made by Nye Bevan [1], that great Welshman and socialist, when in the House of Commons he observed the entry of the Prime Minister while his question was being answered by a subordinate Minister. Bevan brusquely stopped the answer and said

> If we complain about the tune, there is no reason to attack the monkey when the organ grinder is present.

And I was quite convinced that I should, in present circumstances, be regarded simply as the monkey. I was, however, at a loss to know precisely who should be considered as the organ-grinder. Then I realised

---

[1] Aneurin Bevan (1897-1960), British politician, Minister of Health (1945-51) who inaugurated a national health service.

that although I might well be the only monkey, I should be among a multitude of organ-grinders and that this might be the real reason for choosing me. I shared this thought with the organisers and they, to their credit a little shamefacedly, agreed. So here I am.

2. Density Matrix Theory (DMT)

As a graduate student from 1959 until 1962, I was fortunate enough to study with Roy McWeeny and to learn Density Matrix Theory (DMT) from him. He sent me off to a summer school in Sweden run Per-Olov Löwdin where my knowledge of DMT was developed and deepened to the extent that when, a couple of years later, I heard John Coleman speak at Sanibel, I could comprehend the nature of the difficulties with which he was concerned.

From my education at their hands I learned that, for the purposes of understanding molecular theory in the clamped nucleus approach, it was enough to know the two-electron density matrix. Indeed, it was often enough to know this simply in its spinless form. I learned that this two-electron insight was of limited utility because it had not, at that time, proved possible to obtain an equation from which the two-electron density matrix could be directly determined. The only way to get to a correctly formed (N-representable) two-electron density matrix was to start from the full wavefunction for the problem and to integrate out over all but two of the electronic variables.

I actually heard John Coleman speak while I was a Fulbright Fellow in Prof. John C. Slater's Solid State and Molecular Theory Group (SSMTG) at MIT. Although Slater was an extremely supportive of our sub-group, which included Jules Moskowitz, Malcolm Harrison and Imre Csizmadia, in our development of the POLYATOM molecular electronic structure calculation system, he was much less enthusiastic about a density matrix formulation of the electronic structure problem. When I presented him with the first draft of a paper in which the Hartree-Fock equations were expressed directly in density matrix form, to seek his permission for it to be published with the SSMTG address, he, politely but quite firmly, told me to go away and to rewrite it in wavefunction form. He felt that few understood density matrix notation and that because density matrices could not be determined directly, they constituted no advance on a wavefunction approach. Slater was, of course, at that time (1963) much concerned with the X_α method that some regard as the source of the density functional (DFT) approach. I remember forming the opinion that he regarded this approach simply as a reasonable approximation leading to substantial computational savings, but this may be a mistaken perception since I was not at all involved in this side of the work of the SSMTG. I did however feel somewhat put down by Slater's dismissal of DMT and much regretted that I was not able to say to him that he was just a bit out of date and that I'd heard Prof. Coleman's talk at Sanibel in which he had shown that indeed one could get the two-electron density matrix directly. If only I could have said so! But that is what the French call *l' esprit des escaliers*, that which you wish, after the meeting, that you had said during the meeting.

You will, therefore, understand how eagerly I anticipate John Coleman's talk on reduced density matrices. I look forward too, in the same context, to the contribution from Carmela Valdemoro and her coworkers, especially in the hope that it contains a further contribution to the development of a contracted Schrödinger equation (CSE) from which a two-electron density matrix can be directly determined. Perhaps too, Richard Bader's title indicates that he has moved to an even further contraction of the Schrödinger equation. I wonder too, whether the contribution from Roy McWeeny, Harry Quiney and Stephen Wilson might not have some bearing on the possibilities of a contracted scheme but based on a one-particle operators of Dirac rather than Schrödinger form?

I do hope that during the discussion time associated with these papers, and perhaps during other time too, we shall be able to share some insights into the work of Nakutsuji [2], of Yasuda [3], of Mazziotti [4–8] and perhaps also into the cumulant work of Kutzelnigg in this area.

I should like to think that any advances here might properly be offered up to shade of John Slater for, setting aside my experience, it must not be forgotten that in his group flourished both Per-Olov Löwdin and, somewhat later, Roy McWeeny and that Löwdin actually made the first of his contributions on density matrices in a series of papers from MIT.

3. Density Functional Theory (DFT)

After reading Peter Gill's article, *Obituary: Density Functional Theory (1927-1993)* [1], I wondered if I wasn't today going to be involved in a wake rather than anything more jolly. It is interesting that in the obituary he speaks of her (DFT) as the younger sister of wavefunction theory and that he attributes her demise to Hyperparametric Disorder.

By making her a younger sibling of the wavefunction, I suppose he means to imply that she will always be dependent upon him and that attempts at an independent existence, though in principle possible, have led to her downfall and now to her death. It makes it sound a bit like the celebrated *risqué* song (which I don't need to quote in full, I'm sure) that begins

> She was poor but she was honest,
> The victim of a rich man's whim.
> First he loved her then he left her,
> And she lost her honest name.

Could there really be anything in Peter Gill's contention or are the reports of her death, like those of Mark Twain's, "an exaggeration"? What signs of death might be described in the titles of anything offered for this meeting? Perhaps the title of Roy McWeeny's paper does seem to indicate that there are some reasons for concern. Unsolved problems, however, are not necessarily mortal. It would also seem unwise to anticipate too much trouble from inconsistencies in existing functionals for the the Kohn-Sham potential that lead to symmetry problems. But Prof Theophilou will guide us through these matters. Although the title of Prof Levy's contribution on the adiabatic connection does seem to indicate sober and reflective consideration on the patient's condition, it has no hint of anxiety in it. It seems unlikely to me then, that I shall be attending a wake in what is to follow. In fact most of the DFT titles appear very cheerful and seem to anticipate a healthy and active functional, effective not only in some traditional areas of endeavour, exemplified perhaps by the work to be described on the Luttinger liquid and the Mott insulator, but also in some new areas. So there are to be two contributions on time-dependent deployment of DFT. This is somewhere that I had thought DFT unsuited to go and I look forward hearing how it is managing to get there. Similarly I had believed DFT simply could not cope with van der Waals interactions neither could it cope properly with molecular dissociation. But, judging from the title of the talk of Fuchs, Gonze and Burke, I am going to have to reconsider my position here. There are also two talk titles in which the word "correlation" is used. I confess to finding the idea of "correlation" when used in the context of molecular quantum mechanics, an extremely slippery one. Am I dealing with something that has a definite physical meaning? Is there a correlation operator whose expectation value I can, at least in principle, compute? Or am I simply describing a particular characteristic of a model wavefunction/density? I hope to become more enlightened about such matters in the course of the meeting. There seems to be no lack of activity by and for the functional itself. From its title, the talk by Prof. Nagy, would seem to presage a rather general extension of the theory to excited states while the titles of two of the other talks make it seem that functional construction is still a lively interest and, from another title, that the uniqueness of a constructed potential is still an issue, at least in Spin-DFT.

I am really looking forward to hearing what is to be said about combined systems of electrons and nuclei. I know that some recent work on the simultaneous determination of nuclear and electronic wave functions without the Born-Oppenheimer approximation [9] has met with some success, at least for very simple diatomics and I notice that Gidopolous and Kohanoff's talk about electronic structure with quantum nuclei, allows itself the luxury of a wavefunction in the title. But perhaps there was a mistake in my programme draft. I thought that one was expelled from the DFT practioners union for any such titular *bêtises*. There are no such anxieties over Prof Gross's talk, however, which firmly places both electrons and nuclei in a DFT context.

4. Conclusions

When I was an undergraduate I attended, each week or so during my second year, a tutorial on the Philosophy of Science. The tutorial began with me reading my essay on a topic chosen by my tutor. When I had ended, my tutor usually sat silent for a few moments, puffing on his pipe and he would then say "Yes, well, I think that you've covered all the points but ..." and then he'd open up on me. In a curious sort of way I feel now just as I did then but please don't open up on me even if you think I've not covered all the points or perhaps got it wrong. Instead let us go quickly to the real business in hand to learn how to consider properly as many points as we can.

Postscript

What has gone before is what I said at the time. As is perhaps always the case, one might wish some things unsaid. I clearly was rather mixed up about what was involved in some of the work on time dependent approaches and I understand more clearly than I did, just how willing DFT theorists are to accommodate wavefunctions. But it seemed better to leave visible my failures properly to comprehend, for I feel sure that they will have been amply supplied in the articles to follow.

References

[1] Peter M. W. Gill, *Aust. J. Chem* **54**, 661 (2001)
[2] H. Nakatsuji and K. Yasuda, Phys. Rev. Lett. **76**, 1039 (1996)
[3] K. Yasuda and H. Nakatsuji, Phys. Rev. A **56**, 2648 (1997)
[4] D. Mazziotti, Phys. Rev. A **57**, 4219 (1998)
[5] D. Mazziotti, Chem. Phys. Lett. **289** 419 (1998)

[6] D. Mazziotti, Int. J. Quantum. Chem. **70**, 557 (1998)
[7] D. Mazziotti, Phys. Rev. A **60**, 3618 (1999)
[8] D. Mazziotti, Phys. Rev. A **60**, 4396 (1999)
[9] Hiromi Nakai, *Int. journ Quant. chem.* **86**, 511 (2002)

THE PROGRAMME

Editorial Note: We give here the programme as it was presented at registration since this formed the basis of the introductory talk given by Professor Brian Sutcliffe which is recorded in the preceding pages. Subsequent changes to the programme are detailed separately.

Friday, 31st May, 2002

15:30 Registration
    Tea
17:30 Welcome
    S. Wilson
17:40 Forum Preview
    B.T. Sutcliffe
18:00 *Dinner*

Session 1

Chair: R.F.W. Bader

19:30 Perspectives on the Adiabatic Connection and Coordinate Scaling
    M. Levy
20:00 Discussion
20:05 Effective exact exchange Kohn-Sham methods as basis for time-dependent approaches
    A. Goerling and F. Della Sala
20:35 Discussion
20:40 Key concepts in time-dependent current-density functional theory
    R. van Leeuwen, P.L. de Boeij, M. van Faassen, A.J. Berger and

J.G. Snijders
20:10 Discussion
21:15 *Bar*

Saturday, 1st June, 2002

08:00 *Breakfast*

Session 2

Chair: M. Levy

08:45 Density functional theory for the combined system of electrons and nuclei
   E.K.U. Gross
09:15 Discussion
09:20 Correlated Wavefunction Densities as a Tool in Density Functional Theory   D.J. Tozer, G. Menconi, P.J. Wilson, M.J. Allen and T.W. Keal
09:50 Discussion
09:55 Ensemble and subspace excites-state density functional theory
   A. Nagy
10:25 Discussion
10:30 *Coffee*

Session 3

Chair: R. McWeeny

11:00 Reduced density matrices: The Kummer variety
   A.J. Coleman
11:30 Discussion
11:35 Compact forms of Reduced Density Matrices
   L.M. Tel, E Perez-Romero, C Valdemoro and F J Casquero
12:05 Discussion
12:10 Towards van der Waals interactions, chemical accuracy, and proper molecular dissociatlon within density-functional theory: use of the adiabatic-connection fluctuation-dissipation approach

M. Fuchs, X. Gonze and K. Burke
12:40 Discussion
13:00 *Lunch*

Session 4
Chair: E.K.U. Gross

14:00 Some unsolved problems in density functional theory
R. McWeeny
14:30 Discussion
14:35 Symmetry Properties of the Kohn-Sham potential and inconsistencies of existing functionals
A.K. Theophilou
15:05 Discussion
15:10 Density functional theory from the extreme limits of correlation
M. Seidel
15:40 Discussion
15:45 *Tea*

Session 5
Chair: A.K. Theophilou

16:15 Local 'hybrid' functionals based on self-consistent α exchange
V.V. Karasiev and E.V. Ludeña
16:45 Discussion
16:50 A density-functional study of the Luttinger liquid and the Mott insulating phase in the one-dimensional Hubbard model
K. Capelle, N.A. Lima, M.F. Silva and L.N. Oliveira
17:20 Discussion
17:25 Investigation of non-local XC functionals within Density Functional Theory
S.J. Clark, D J Tozer and P P Rushton
17:55 Discussion
18:00 A new wave function method for electronic structure calculations with quantum nuclei
J. Kohanoff and N. Gidopoulos
18:30 Discussion

19:00 *Dinner*

Session 6
Chair: B.T. Sutcliffe
20:30 Posters

Sunday, 2nd June, 2002.

08:00 *Breakfast*

Session 7

Chair: E.V. Ludena

09:00 Using the physics of an open system to extract chemistry from the one-electron density matrix: The only matrix chemistry requires
R.F.W. Bader

09:30 Discussion

09:35 On the Uniqueness of Potentials in Diagonal Spin Density Functional Theory
N. Gidopoulos

10:05 Discussion

10:10 The relativistic strong orthogonality condition
R McWeeny, H M Quiney and S Wilson

10:40 Discussion

10:45 *Coffee*

Session 8

Chair: B.T. Sutcliffe

11:00 The Forum
12:45 Close
13:00 *Lunch*

Posters

Scaling the spin densities separately in density functional theory

R.J. Magyar, T.K. Whittingham and K. Burke

Some new applications for plane-wave DFT calculations in chemistry
Carole A. Morrison and Murshed M. Siddick

Calculation of INS (inelastic neutron scattering) Spectra using DFT methods
A.J. Ramirez-Cuesta and J. Tomkinson

Some Basic Properties of the Correlation Matrices
D.R. Alcoba, C. Valdemoro, L.M. Tel, E. Perez-Romero and F.J. Casquero

Development of nonlocal exchange-correlation functionals
P.P. Rushton, D.J. Tozer and S.J. Clark

Title to be announced
J.J. Lee

Programme changes

Saturday, 1st June, 2002

Session 5

16:15 The problem of functional N-representability in density matrix and density functional theory. Illustrations using Hooke's atom
E.V. Ludena

16:45 Discussion

Session 6

The following papers were given as posters:-

On the Uniqueness of Potentials in Diagonal Spin Density Functional Theory
N. Gidopoulos

The relativistic strong orthogonality condition
R. McWeeny, H.M. Quiney and S. Wilson

Sunday, 2nd June, 2002.

Session 7

10:10 Local 'hybrid' functionals based on self-consistent α exchange
V.V. Karasiev and E.V. Ludena

10:40 Discussion

ABSTRACTS OF TALKS AND POSTERS

The Fundamentals of Electron Density, Density Matrix and Density Functional Theory for Atoms, Molecules and the Solid State - A Forum Preview
B.T. Sutcliffe,
*Laboratoire de Chimie Physique Moléculaire,
Université Libre de Bruxelles, B-1050 Bruxelles, Belgium*

The topics to be discussed at this meeting are surveyed and an attempt is made to provide a context in which they might be placed.

Perspectives on the Adiabatic Connection and Co-ordinate Scaling
M.P. Levy
*Department of Chemistry,
Tulane University,
New Orleans, Louisiana 70188, U.S.A.*

Effective exact exchange KS methods as basis for time-dependent DFT and multireference CI approaches
A. Görling, F. Della Sala,
*Lehrstuhl für Theoretische Chemie,
TU München, D-85747, Garching, Germany*

An effective exact exchange Kohn-Sham method, the localized Hartree-Fock (LHF) method, is presented. The asymptotic behavior of the Kohn-Sham exchange potential is reconsidered and shown to exhibit an unexpected behavior on nodal surfaces of the energetically highest occupied orbital. The LHF exchange potential is shown to have the same asymptotic behavior as the exact exchange potential. The performance of LHF orbitals in time-dependent density- functional methods and in multireference CI procedures is discussed.

References

F. Della Sala and A. Görling, J. Chem. Phys. 115, 5718 (2001); 116, 5374 (2002).

T. Hupp, B. Engels, F. Della Sala, and A Görling, Chem. Phys. Lett., in press.

Key concepts in time-dependent current-density functional theory

Robert van Leeuwen, P.L.de Boeij, M.van Faassen, A.J.Berger, J.G.Snijders

Theoretical Chemistry, Materials Science Centre, Rijksuniversiteit Groningen, Nijenborgh 4 9747 AG, Groningen, The Netherlands

We describe some advantages of using the current-density as a basic variable in density functional theory, especially with regards to non-local polarization phenomena. The fundamental equations of current-density functional theory are derived from an action principle using the time-contour formulation of Keldysh. Some ways of constructing the exchange-correlation part of the vector potential are given as well as some directions for future progress.

Density functional theory for the combined system of electrons and nuclei

E.K.U. Gross

Institut für Theoretische Physik, Freie Universität Berlin, Arnimallee 14, 14195 Berlin, Germany

Traditional density functional theory, necessarily and inevitably, involves the Born-Oppenheimer approximation: One calculates the electronic ground-state density corresponding to the electrostatic potential generated by clamped nuclei. There are, however, many situations where the coupling between the electronic and the nuclear motion is important. Prime examples are the branching ratios of chemical reactions, the electron-phonon interaction with superconductivity as its most dramatic consequence, and the laser-induced dissociation of molecules. The subject of this lecture is to describe how one can go beyond the Born-Oppenheimer approximation by treating the fully coupled system of electrons and nuclei in terms of a multi-component density functional theory. After setting up the formal framework, the construction of approximate functionals for the electron-nuclear correlation energy will be described [1]. First results for molecular ground-state properties will be presented. Furthermore, a time-dependent generalization will be developed to treat molecules in strong laser pulses. Finally some results for conventional (phonon-mediated) superconductors will be presented.

Reference
[1] T. Kreibich and E.K.U. Gross, Phys. Rev. Lett. **86**, 2984 (2001).

Correlated wavefunction densities as a tool in DFT
David J Tozer and Mark J Allen
Department of Chemistry, University of Durham, UK.

The dispersion interaction in the helium dimer is considered from the viewpoint of the force on a nucleus. By considering correlated electron densities, the atomic density distortion associated with the dispersion force is highlighted. The correlated densities are then used to determine high quality correlation potentials within the Hartree-Fock-Kohn-Sham (HFKS) formalism. HFKS calculations are performed using these potentials and the associated forces are compared with reference dispersion forces. The partitioning of the potential into atomic and interaction components is investigated.

Density matrix functional theory
Á. Nagy
Department of Theoretical Physics, University of Debrecen, H-4010 Debrecen, Hungary

Recently there has been a growing interest in the theory of density matri- ces. The theory goes back to the pionering works of Husimi [1] and Lowdin [2]. The problem of N-representability [3] turned out to be a serious difficulty that hin- dered the theory of becoming a powerful tool for treating many-electron systems. A breakthrough has come with the recent research of Nakatsuji [4], Valdemoro [5] and Mazziotti [6]. The NVM theory presents an alternative to traditional many-body quantum calculations through solving the contracted Schrodinger equation.

In this talk another approach is presented. Recent results of Gonis et al. [7] is reviewed and generalized. They extended the Hohenberg-Kohn theorems [9] showing that the total energy is a functional of the 'hyperspace density' that is the diagonal of the spin independent reduced second-order density matrix. Another approach based on the pair density, i.e. a pair density functional theory was proposed by Ziesche [8].

Auxiliary equations are derived through adiabatic connection [10]. It is shown that in the ground state the diagonal of the spin independent second-order density matrix n can be determined by solving a single auxiliary equation of a two-particle problem. Thus the problem of an arbitrary system with even electrons can be reduced to a two-particle problem. The effective potential of the two-particle equation contains a term of completely kinetic origin.

References

[1] K. Husimi, Proc. Phys. Math. Soc. Jpn. 22, 264 (1940).
[2] P O Lowdin, Phys. Rev. 97, 1474 (1955).
[3] A. J. Coleman, Rev. Mod. Phys. 35, 668 (1963); E. R. Davidson, Reduced Density Ma- trices in Quantum Chemistry (Academic Press, New York, 1976); A. J. Coleman and V. I. Yukalov, Reduced Density Matrices: Coulson's Challange (Sringer-Verlag, New York, 2000); J. Cioslowski, Many-electron Densities and Reduced Density Matrices (Kluwer/Plenum, New York, 2000).
[4] H. Nakatsuji and K. Yasuda, Phys. Rev. Lett. 76, 1039 (1996); K. Yasuda and H. Nakatsuji, Phys. Rev. A 56, 2648 (1997).
[5] F. Colmenero and C. Va demoro, Phys. Rev. A 47, 979 (1993); C. Valdemoro, L. M. Tel and E. Perez-Romero, Adv. Quantum. Chem. 28 33 (1997).
[6] D. Mazziotti, Phys. Rev. A 57, 4219 (1998 ; Chem. Phys. Lett. 289 419 (1998); Int. J. Quantum. Chem. 70 557 (1998); Phys. Rev. A 60, 3618 (1999); 60, 4396 (1999).
[7] A. Gonis, T. C. Schuithess, J. van Ek and P. E. A. Turchi, Phys. Rev. Lett. 77, 2981 (1996); A. Gonis, T. C. Schulthess, P. E. A. Turchi and J. van Ek, Phys. Rev. B 56, 9335 (1997).
[8] P. Ziesche, Phys. Lett. A 195, 213 (1994); Int. J. Quantum. Chem. 60, 149 (1996); M. Levy and P. Ziesche, J. Chem. Phys. 115, 9110 (2001).
[9] P. Hohenberg and W. Kohn, Phys. Rev. 136, B864 (1964)
[10] A Nagy, Phys. Rev. A (in press).

Reduced Density Matrices: The Kummer Variety
A. J. Coleman
Department of Mathematics and Statistics, Queen's University, Kingston, Ontario, Canada

The talk will, essentially, be an introduction to a paper to appear in Physical Review A entitled "The Kummer Variety, the Geometry of N-Representability and Phase Transitions". We show that the energy of a system of identical fermions or bosons, for which the Hamiltonian involves at most two-particle interactions, and that most of the properties of such systems, of current interest in physics and chemistry, follow from a knowledge of the Second Order Reduced Density Matrix. That is, the 1959 conjecture of Charles Coulson is correct: theoretical physics and chemistry is possible without explicit knowledge of Wave Functions.

Compact forms of Reduced Density Matrices

L. M. Tel, C. Valdemoro, E. Perez-Romero, F. J. Casquero

Facultad de Quimica, Universidad de Salamanca, 37008 Salamanca, Spain

The compact forms of the RDMs are matrices with many vanishing elements, which nevertheless convey the same information as the RDMs themselves. The RDMs can be obtained from their compact forms by projection. The minimum number of non-zero elements is easily calculated. Thus, the compact forms can be used to save storage space, as unique variational parameters, or as safe and economical objects when improving the approximation of an RDM.

Towards van der Waals interactions, chemical accuracy, and proper molecular dissociation within density-functional theory: use of the adiabatic-connection fluctuation-dissipation approach

Martin Fuchs,[*,1] Xavier Gonze,[1] and Kieron Burke[2]

[1] *Unité de Physico-Chimie et de Physique des Matériaux, Universite Catholique de Louvain, 1348 Louvain-la-Neuve, Belgium*

[2] *Department of Chemistry and Chemical Biology, Rutgers University, Piscataway, NJ 08854, U.S.A.*

Density-functional theory (DFT) has been highly valuable in analyzing the ground state properties of complex electronic systems, be they molecular, solid state, or surface systems. A cornerstone for this success is the availability of accurate and robust approximations for the unkown exchange- correlation functional. Traditional schemes such as the local-density (LDA) or generalized-gradient approximations (GGA) often achieve a remarkable accuracy. Yet more sophisticated approximations are needed to (1) describe the energetics of chemical reactions and the related potential energy surfaces with "chemical accuracy" and (2) properly treat the van der Waals attraction between non-overlapping systems. Systematic progress beyond today's exchange-correlation functionals is expected from the (in principle exact) representation of the exchange-correlation energy through the adiabatic-connection fluctuation-dissipation theorem (ACFDT). There the electronic pair-correlations are obtained by a coupling-strength integration of the dynamical density response that may be computed in the framework of time-dependent DFT. We have implemented the ACDFT formalism within a pseudopotential plane-wave method,[1,2] using the random phase approximation (RPA) together with a local-density correction for short- range correlations, and an (approximate) exchange kernel. As an application, we have calculated the (groundstate) potential energy curves for the H_2 and Be_2 dimers. We find that the RPA gives excellent results for the binding energy, equilibrium bond length and fundamental vibrational mode in these dimers.[1] For stretched Be_2, the RPA yields the expected van der Waals attraction, unlike the LDA or GGA's. For stretched H_2, the strong static correlations that arise when the dimer dissociates into the free atoms are largely recovered by the RPA,[3] without the need for (symmetry breaking) spin-orbitals as in LDA or GGA. Finally, some links to alternative methods and open questions will be presented. Our work shows that ACFDT exchange-correlation functionals are now computationally feasi- ble and are promising candiates for an accurate, seamless calculation of atomic potential energy surfaces all the way from the bonding to the van der Waals regime.

* Present address: Fritz-Haber-Institut der MPG, Faradayweg 4-6, 14195 Berlin, Germany

References

[1] M. Fuchs and X. Gonze, Phys. Rev. B, in press.
[2] X, Gonze et al., Comput. Materials Science. See also http://www.abinit.org.
[3] M. Fuchs, K. Burke, Y.-M. Niquet, and X. Gonze, submitted.

Some Unsolved Problems in DMT and DFT

Roy McWeeny,
Dipartimento di Chimica, Universita di Pisa,
Via Risorgimento, 35, 86100 Pisa, Italy

Density functional theory (DFT) has its roots in density matrix theory (DMT). This contribution therefore opens with a brief survey of the 1- and 2-electron density matrices (1-DM and 2-DM) and related densities (charge, spin, current, and the pair density), – together with their main properties.

The problem of how to pass from the DMs to an energy density functional is considered in the light of simple explicit examples: these show the richness of structure hidden within the 1-DM and 2-DM – even in an IPM approximation. In DFT, the diagonal element of the spinless 1-DM (i.e. the electron density, P) is deemed sufficient to determine everything else, including the 2-DM and the related electron interaction energy. In DMT, the 1-DM and 2-DM are rigorously sufficient. But in either case the densities must be subject to the condition that they derive from an acceptable N-electron wave function – they must be *N-representable*. If we disregard this condition, using assumed densities and trying to optimize the energy, then variational 'collapse' is inevitable and the results will be meaningless.

In the absence of an acceptable procedure for imposing N-representability, it is customary to use an *ansatz* for the densities – forms which are *known* to arise (by integration) from an N-electron function of some suitable approximate form. Thus, for example, an IPM system with a 1-determinant wave function permits the calculation of *all* the DMs (for $1-, 2-, ...N$-electron clusters) in terms of the Fock-Dirac 1-DM. More generally, a CI-type wavefunction leads to a 'natural-orbital' expansion of similar form (but with non-integral occupation numbers) for the 1-DM – but no simple form for even the 2-DM.

Alternative procedures for getting the DMs are studied. They are essentially group theoretical and involve the separation of space and spin variables: one starts from the the Schrödinger equation $H\Phi = E\Phi$, which in non-relativistic approximation does not make any reference to spin. This could in principle be solved exactly, but then we must ask which of the solutions are physically admissible (most of them will not be) and what constraints they impose on the density matrices. For example, for a 10-electron system in a singlet state, determination of the DMs requires knowledge of 42 linearly independent spatial eigenfunctions.

The absolute minimum requirement for eliminating all reference to the N-electron wave function appears to be the availability of an N-representable spinless 2-DM.

Symmetry Properties of the Kohn and Sham potential and its relation to the external potential.

Andreas K. Theophilou,
"DEMOCRITOS" National Center for Scientific Research,
Institute of Materials Science
TT 15310, Athens, Greece

A fundamental problem of DFT is the determination of the Kohn potential $V_K(\mathbf{r}, \rho)$ (Hartree plus Exchange and correlation potential) as a function of the density. A necessary condition for this potential is invariance under the symmetry group G of the external potential, as otherwise no eigenstates of the noninteracting Hamiltonian, transforming according to the irreducible representations of the symmetry group G of the exact Hamiltonian, exist. Unfortunately the density does not have in general the symmetry properties of the external potential and therefore all existing general expressions for $V_K(\mathbf{r}, \rho)$ do not comply with the above necessary condition. This inconsistency implies that all $V_K(\mathbf{r}, \rho)$ have limited accuracy. For this reason we decided to express this potential as a function of the external potential, choosing properly its symmetry and asymptotic properties. The explicit forms obtained are used for the determination of the ground state energies of atoms with nuclear charge varying from 2 to 20. As we have no exact expression for the energy as a functional of the potential, we used the Hartree-Fock expression. The parameters of the Kohn potential were varied so as to obtain the minimum of the energy. The deviation of the energies calculated from the experimental ones are less than 0.05%. It is to be noted that our present methodology resembles more to the Optimised Effective Potential Approximation. We hope that by using higher order approximations for $V_K(r, V)$ we shall be able to improve the accuracy of DFT. Then the advantage in relation to conventional DFT will be significant, as given the exact external potential the Kohn potential will be determined automatically, and the speed of calculations will increase dramatically as no iterations for self-consistency will be needed.

Approach to density functional theory starting from the extreme limits of correlation

Michael Seidl and Markus Pindl

University of Regensburg,
D-93040 Regensburg,
Germany)

For most electron systems the perturbation expansion of the correlation energy, starting from the weak-interaction limit, is divergent. Nevertheless, its low-order terms provide valuable information if a proper re-summation is performed.

In density functional theory, the limit of strong repulsion between the electrons in a given density distribution is mathematically simple. In particular, this strong-interaction limit allows to extrapolate accurately the incomplete results of the perturbation expansion around the weak-interaction limit.

In order to investigate the reason for the finite radius of convergence of that expansion we also study the regime of "attractive electrons" ("negative repulsion") and find a simple estimate for the second-order correlation energy.

Local "hybrid" functionals based on exact-expression approximate exchange

Valentin V. Karasiev

Centro de Química, Instituto Venezolano de Investigaciones
Científicas, IVIC Apartado 21827, Caracas 1020-A, Venezuela

We propose to replace the nonlocal Hartree-Fock exchange in conventional exchange-correlation hybrid functionals by the local exact-expression approximate exchange (EEAX) which is the self-consistent α (SCα) [1-3], the asymptotically-adjusted self-consistent α (AASCα) [4], the Localized Hartree-Fock (LHF) method introduced by Görling [5] or the approximation to the optimized effective potential (OEP) method introduced by Krieger, Li, and Iafrate [6]. Such replacement is justified by the fact that the exchange energy expression of the EEAX functionals is equal to the Hartree-Fock one (exact exchange) while the corresponding EEAX exchange potential is a local, multiplicative operator. Applications to diatomic molecules for the case of EEAX=SCα, AASCα are presented. Performance of new totally local "hybrid" exchange-correlation functionals is quite close to that of original non-local hybrid models.

References

[1] V. Karasiev, E. V. Ludeña, R. López-Boada, Int. J. Quantum Chem. 70, 591 (1998).
[2] V. V. Karasiev, J. Mol. Struct. (THEOCHEM), 493, 21 (1999).
[3] V. V. Karasiev and E. V. Ludeña, Phys. Rev. A 0625XX (2002) (in press).
[4] V. V. Karasiev and E. V. Ludeña, Phys. Rev. A 032515(1-8) (2002).
[5] F. Della Sala, and A. Görling, J. Chem. Phys. 115, 5718 (2001).
[6] J. B. Krieger, Yan Li, and G. J. Iafrate, Phys. Rev. A 45, 101 (1992).

Density-functional theory for the Hubbard model: numerical results for the Luttinger liquid and the Mott Insulator

K. Capelle

Departamento de Química e Física Molecular Instituto de Quimica de São Carlos, Universidade de São Paulo, Caixa Postal 780, São Carlos, 13560-970 SP, Brazil

N. A. Lima, M.F. Silva, and L. N. Oliveira

Departamento de Física e Informatica Instituto de Física de São Carlos, Universidade de São Paulo, Caixa Postal 369, 13560-970 São Carlos, SP, Brazil

We construct and apply an exchange-correlation functional for the one-dimensional Hubbard model. This functional has built into it the Luttinger-liquid and Mott-insulator correlations, present in the Hubbard model, in the same way in which the usual ab irlitio local-density approximation (LDA) has built into it the Fermi-liquid correlations present in the electron gas. An accurate expression for the exchange-correlation energy of the homogeneous Hubbard model, based on the Bethe Ansatz (BA), is given and the resulting LDA functional is applied to a variety of inhomogeneous Hubbard models. These include finite-size Hubbard chains and rings, various types of impurities in the Hubbard model, spin-density waves, and Mott insulators. For small systems, for which numerically exact diagonalization is feasible, we compare the results obtained from our BA-LDA with the exact ones, finding very satisfactory agreement. In the opposite limit, large and complex systems, the BA-LDA allows to investigate systems and parameter regimes that are inaccessible by traditional methods.

Investigation of non-local XC functionals within density funtional theory.

S.J. Clark[1], D. J. Tozer[2] and P. P. Rushton[1]

[1]*Department of Physics, University of Durham, South Road, Durham DH1 3LE, UK*

[2]*Department of Chemistry, University of Durham, South Road, Durham DH1 3LE, UK*

The non-local weighted density approximation (WDA) was first developed over twenty years ago but has been left in relative obscurity in favour of the less computationally demanding local density and generalised gradient approxima- tions. As such, the functional form is underdeveloped and appears to gives struc- tural parameters, *e.g.* unit cell sizes, with a similar accuracy to the LDA/GGA. Here we investigate various functional forms of the WDA and its properties for a broad range of real and model systems. Firstly we examine inhomogeneous electron gas densities that are confined in three dimensions and also along one dimension, using a model external potential. We also investigate a range of pair-correlation functions for the WDA by calculating the lattice constant, bulk modulus and bandstructure of Si. Exchange-correlation holes are also determined at specific points in the [110] plane of Si and a comparison is made with data from variational Monte Carlo (VMC) simulations. We show that the WDA is particularly sensitive to the pair-correlation function used and that the accuracy of the lattice constant and bulk modulus is directly related to the quality of the exchange-correlation hole in high density regions. The form of the pair-correlation function also has a significant effect on the electronic band gap.

We also present exchange-correlation energy densities e_{xc} total energies E_{xc} and holes, calculated for strongly inhomogeneous electron gases using the WDA and show that the results closely resemble variational Monte Carlo simulations performed recently, demonstrating the effectiveness of a non-local density functional description.

The One-Electron Density Matrix: The Only Matrix Chemistry Requires

R.F.W. Bader,
*Department of Chemistry,
McMaster University,
Hamilton Ontario, L8S 4M1,
Canada*

The underlying operational observation of chemistry - that of a functional group exhibiting a characteristic set of properties - requires that the density distribution of a group and hence its properties, be relatively insensitive to changes in its neighbouring groups. However, the fields that are used in the determination of the wave function and of the density in DFT are not short-range, but instead reflect the long-range nature associated with the individual e-n and e-e Coulombic fields. The observation upon which the theory of atoms in molecules is founded concerns the paralleling transferability of the electron density with all 'dressed' property densities. The virial field $V(\mathbf{r})$, the virial of the Ehrenfest force exerted on an electron, describes the energy of interaction of an electron at some position $\mathbf{r}$ with all of the other particles in the system, averaged over the motions of the remaining electrons. When integrated over all space it yields the total potential energy of the molecule, including the nuclear energy of repulsion. Because of this inclusion, $V(\mathbf{r})$ yields the most short-range description possible of the potential interactions in a many-electron system. V(r), as well as the kinetic energy density $G(\mathbf{r})$ and their sum $E_e(\mathbf{r})$, all demonstrably parallel the short-range behaviour underlying the transferable nature of $D(\mathbf{r})$. All three energy fields are determined by the one-matrix whose diagonal elements in addition determine $D(\mathbf{r})$. Thus it is the one-matrix that is short-ranged and responsible for the observation of functional groups with characteristic, transferable properties in chemistry. This talk describes a method for incorporating the virial field into the self-consistent field calculation to obtain an exact prescription of the 'average field' experienced by a single electron in a many-electron system.

A new wavefunction method for electronic structure calculations with quantum nuclei

Jorge Kohanoff

Queen's University, Belfast, Northern Ireland

Nikitas Gidopoulos,

ISIS Science Theory Division, Rutherford Appleton Laboratory, Oxfordshire, England

Electronic structure calculations within DFT are normally carried out in the clamped nuclei (classical nuclei) approximation. While this is a sensible idea for many systems of interest, the quantum mechanical treatment of nuclear degrees of freedom is desirable in a wide variety of situations. These range from molecules involving light-mass atoms like H, or even first-row elements, to liquids like water, and to solids like hydrogen, ice and KDP.

To include the influence of quantum nuclear delocalization on the electrons it is needed to solve the nuclear Schroedinger equation in the multidimensional *ab initio* (DFT or another) potential energy surface (PES). Two problems arise at this point: 1) solving the 3N-dimensional nuclear Schroedinger equation (where N is the number of atoms), and 2) mapping the 3N-dimensional PES. While the first limitation can be overcome by a suitable approximate nuclear wave function, the mapping itself is a formidable task. It is then desirable to find alternatives to this approach. One such alternative is the so-called ab initio path integral method, which has given very interesting fruits in the past decade. An alternative approach has been proposed by Kreibich and Gross, in terms of a multi-component DFT. Kohn-Sham equations can be derived and solved and, moreover, these can be extended to the time-dependent case.

In this work we have taken an alternative road, focusing for the moment being in the ground state. The method employs the Born-Oppenheimer (BO) separation of the total wavefunction as a product of an electronic times a nuclear wavefunction, and an additional product, or, Hartree ansatz for the nuclear wavefunction. The quality of the approximation depends on the coordinates in which the nuclear wavefunction is represented. For the original nuclear coordinates, it is rather bad, but it becomes an excellent approximation for the normal coordinates, because these are naturally quite uncorrelated. The method proceeds by minimizing the Hartree energy with respect to both, single-coordinate wavefunctions and the coordinates themselves. The problem of sampling the multidimensional PES is still present, because the potential felt by each mode is averaged over all the remaining modes. However, the decomposition into the least correlated modes allows us to approximate this average by a low-order Taylor expansion of the (N) single-mode potential surfaces. The zero-order approximation is obtained by replacing the (N-1)- remaining single-mode wavefunctions with a delta-functions centered at the mean value. The second order approximation incorporates information about the coupling of this coordinate to the remaining (N-1). In this way we determine accurately the nuclear wavefunction using knowledge of a very small portion of the PES, and thus reducing greatly the computational effort. The part of the PES that has to be calculated is not known a priori, but has to be determined self-consistently on-the-fly, e.g. from ab initio (DFT) calculations.

By testing the accuracy of our approximation against the full BO solution in the case of triatomic molecules, for which the full BO hypersurface is obtainable, we find excellent agreement between the two theories, while an inappropriate choice of internal coordinates results in an inaccurate wavefunction. Moreover, we show that the second order approximation is essentially equivalent to full Hartree, but at a massively lower and scalable cost, while the order zero is also an excellent approximation.

On the Uniqueness of Potentials in Diagonal Spin Density Functional Theory
N. Gidopoulos,
ISIS Science Theory Division, Rutherford Appleton Laboratory, Chilton, Didcot, Oxfordshire, England

We observe that a small constant shift of the magnetic field in the class of Hamiltonians which SDFT addresses, is trivial, in the sense that it does not alter the eigenstates or change their orden An immediate consequence is that the magnetic field in SDFT should not be expected to be determined absolutely but only within a critical constant, which is the energy to flip a spin. This observation resolves the non-uniqueness in the determination of spin-potentials in SDFT discovered recently by Capelle and Vignale. We further show that the freedom of a constant shift in the magnetic field is the only freedom availabie; otherwise, different spin potentials, i.e. potentials which differ by more than a spin-dependent constant, cannot have a common eigenstate. Hence, the invertibility of the mapping between spin-potentials and densities is established.

Finally, we see how the formulation of SDFT can be modified to accommodate the property that spin potentials are determined only within this spin-constant, by constraining the spin density to integrate to a fixed number of spin up and spin down particles. In consequence, density functionals in SDFT must also depend implicitly on these numbers. At the end, the number of spin up and down particles for the ground state are always determined variationally by considering the lowest ground state energy of systems constrained to have fixed numbers of spin up and down electrons.

Relativistic density matrices: The relativistic strong orthogonality condition

R. McWeeny,
Dipartimento di Chimica, Universita di Pisa,
Via Risorgimento, 35, 86100 Pisa, Italy

H.M. Quiney,
School of Chemistry, University of Melbourne,
Parkville, Victoria 3010, Australia

S. Wilson,
Rutherford Appleton Laboratory,
Chilton, Oxfordshire OX11 0QX, England.

The relativistic strong orthogonality condition is considered within the Furry-bound state interaction picture of quantum electrodynamics which provides the foundation of molecular electronic structure theory when the effects of special relativity are significant. Strong orthogonality is an important ingredient of the group function model. In the relativistic group function model the number of electrons associated with each group is no longer constant since virtual electron-positron pair processes may occur. The separation of single particle state functions into mutually disjoint sets is shown to be a necessary and sufficient condition for strong orthogonality. In general, these mutually disjoint sets may contain both positive energy and negative energy single particle state functions. The no-virtual-pair approximation emerges as a special case.

Some new applications for plane wave DFT calculations in chemistry

C. A. Morrison,[a] M. M. Siddick[ab]
[a]*Department of Chemistry,* [b]*Department of Physics, University of Edinburgh, Edinburgh, Scotland*

The overwhelming majority of work currently undertaken in quantum mechanical computational chemistry is based on codes that embrace an isolated molecule approach; that is, the system of interest is alone in an infinite, empty universe. Whilst this style of modelling is clearly relevant to the study of gas-phase systems, it is not so nearly applicable to the condensed state. Many interesting questions on structure and bonding could be addressed if a modelling approach capable of handling a periodic lattice of interacting molecules, such as plane wave DFT, were adopted.

In our research group we have recently developed two new applications for plane-wave DFT calculations for molecular systems. The first tackles the problem of crystal structure disorder, a cause of frustration commonly encountered in X-ray diffraction where more than one model is present in the asymmetric unit of the crystallographic cell.[1] Such structures often suffer from low precision and fine structural detail is lost. Our second application involves the study of intermolecular interactions in a periodic lattice; results obtained for test cases ammonia and urea form the basis of this poster presentation.

Reference
[1] C. A. Morrison, D. S. Wright and R. A. Layfield Interpreting Molecular Crystal Disorder in Plumbocene, Pb(C5H5)2: Insight from Theory, J. Am. Chem. Soc., in press.

Some Basic Properties of the Correlation Matrices
C. Valdemoro[1], D. R. Alcoba[1], L. M. Tel[2]
[1]Instituto de Matematicas y Física Fundamental,
CSIC, Serrano 123,
28006 Madrid, Spain
[2]Departamento de Química Física,
Universidad de Salamanca,
37008 Salamanca, Spain

The p-body Correlation Matrices, obtained by decomposition of the p-order Reduced Density Matrices have very important properties. Amongst the results reported here, the most relevant one is the demonstration that the set of all the first-order transition Reduced Density Matrices of a system provides a complete information about that system. Also, the inter-relation between the properties of the 2-body Correlation Matrix and the second-order Reduced Density Matrix N-representability conditions is studied. Thus, a new procedure aimed at controlling the N-representability deffect of an approximated second-order Reduced Density Matrix is reported. This procedure should prove to be very effective in enhancing the convergence of the iterative solution of the contracted Schrödinger equation besides being of direct applicability in other methods which are Reduced Density Matrices oriented.

Investigations of a non-local exchange-correlation functional
Philip Rushton[1], D.J. Tozer[2] and S.J. Clark[1]
[1]Department of Physics, University of Durham, South Road, Durham
[2]Department of Chemistry, University of Durham, South Road, Durham

The non-local weighted density approximation (WDA) is analysed for strongly inhomogeneous jellium systems and shown to be in very good agreement with recent quantum Monte Carlo simulations. The WDA also provides a good description of quasi two- dimensional (2D) jellium systems, in comparison to conventional LDA/GGA functionals which breakdown in the strong quasi-2D limit.

Charge density and topological analyses of an iron nitrosyl compound

J. J. Lee[1,2], J.A.K.Howard[2] and Yu Wang[1]

[1] *Department of Chemistry, National Taiwan University, Taipei, Taiwan*

[2] *Department of Chemistry, Durham University, Durham, UK*

Charge density and bond characterization are investigated on Iron- nitrosyl dithiocarbamate complex, [Fe $(NO)(S_2CNC_2H_6)_2$] in terms of accurate single crystal diffraction at 100K and an open-shell DFT calculation. The complex crystallizes in the space group of P 2_1/n. The iron atom is in a five coordinated environment with four sulfur and one nitrogen atoms in a square pyramid fashion, the site symmetry is roughly C_{2v}. The iron atom is 0.6(1) Åabove the plane of four sulfur atoms, the nitrosyl group (NO) is in the axial direction perpendicular to the plane. Unfortunately, the oxygen atom has exhibit positional disorder. The Fe-N-O degree is 171.02(6). Since the complex is related to a potential free radical scavenger, this study may shine some light on how it is so via electron density distribution, specially at the iron site and at the NO group. It is known that NO group could be a radical neutral species or NO^+, a nitrosyl, or NO^-, a nitroxide group. Thus the unpair electron can be located either at Fe or at NO group. According to the electron density distribution based on the multipole model and on the DFT calculation, the electronic configuration of iron atom is $4s^1 3d^6$, therefore a Fe(I) and NO group is a nitrosyl group NO^+. In other words, the unpair electron is at Fe not NO group, which is consistent with the conclusion made out of a single crystal EPR and a Moessbauer measurements. Topological analysis on the total electron density will give the bond characterization in terms of topological properties associated with bond critical points. The VSCC of Fe in this five coordinated complex is quite interesting, the detail description and its correlation to the metal ligand bond will be presented.

Scaling the spin densities separately in density functional theory

R J Magyar, T K Whittingham, K Burke

Rutgers Physics Department, currently at the FU Berlin, Germany

Co-ordinate scaling of each spin density separately is considered in spin density functional theory. A virial theorem relates the spin-scaled correlation energy to the spin-scaled correlation potentials. An adiabatic connection formula expresses energies at different spin interaction strengths in terms of spin scaling. Several popular approximate functionals are evaluated on the spin-scaled densities of atoms and of the uniform electron gas. The differences between this and uniform scaling are discussed.

LIST OF PARTICIPANTS

D.R. Alcoba,
Insituto de Matemáticas y Física Fundamental,
CSIC,
Serrano 123,
28006 Madrid,
SPAIN
e-mail: ceer323@imaff.cfmac.csic.es

R.F.W. Bader,
Department of Chemistry,
McMaster University,
Hamilton,
Ontario,
L8S 4M1,
CANADA
e-mail: bader@mcmaster.ca

O. Beckstein,
Department of Biochemistry,
University of Oxford,
South Parks Road,
Oxford,
OX1 3QU,
ENGLAND
e-mail: oliver@biop.ox.ac.uk

K. Capelle,
Departamento de Química e Física Molecular,
Instituto de Química de Sao Carlos,

Universidade de Sao Carlos,
Caixa Postal 780,
Sao Carlos,
13560-970,
BRAZIL

S.J. Clark,
Department of Physics,
University of Durham,
Science Laboratories,
South Road,
Durham DH1 3LE,
ENGLAND
e-mail: sj.clark@durham.ac.uk

A.J. Coleman,
Department of Mathematics and Statistics,
Queen's University,
CANADA
e-mail: colemana@post.queensu.ca

C. Domene,
Department of Biochemistry'
University of Oxford,
South Parks Road,
Oxford,
OX1 3QU,
ENGLAND

M. Fuchs,
Unité de Physico-Chimie et de Physique des Matériaux,
Université Catholique de Louvain,
1348 Louvain-l-Neuve,
BELGIUM

Present address:-
Fritz-Haber-Institut der MPG,
Faradayweg 4-6,
14195 Berlin,
GERMANY
e-mail: fuchs@fhi-berlin.mpg.de

N.I. Gidopoulos,
ISIS Science Theory Division,
Rutherford Appleton Laboratory,
Chilton,
Didcot,
Oxfordshire,
OX11 0QX,
ENGLAND
e-mail: N.Gidopoulos@rl.ac.uk

A. Görling,
Lehrstuhl für Theoretische Chemie,
TU München,
D-85747 Garching,
GERMANY
e-mail: goerling@theochem.tu-muenchen.de

E.K.U. Gross,
Insitut für Theoreische Physik,
Freie Universität Berlin,
Arnimallee 14,
14195 Berlin,
GERMANY
e-mail: hardy@physik.fu-berlin.de

P.J. Grout,
Physical and Theoretical Chemistry Laboratory,
University of Oxford,
South Parks Road,
Oxford OX1 3QZ,
ENGLAND
e-mail: peter.grout@chemistry.oxford.ac.uk

V.V. Karasiev,
Centro de Química Instituto Venezolano de Investigaciones Científicas,
IVIC,
Apartado 21827,
Caracas 1020-A,
VENEZUELA

P.J. Knowles,
School of Chemistry,
University of Birmingham,
Edgbaston,
Birmingham, B15 2TT,
ENGLAND
e-mail: PJ.Knowles@bham.ac.uk

J. Kohanoff,
Atomistic Simulation Group,
School of Mathematics and Physics,
The Queen's University of Belfast,
Belfast BT7 1NN,
Northern Ireland
e-mail: j.kohanoff@qub.ac.uk

J.J. Lee, Department of Chemistry,
National Taiwan University,
Taipei,
TAIWAN
and
Department of Chemistry,
Durham University,
Durham,
ENGLAND

M.P. Levy,
Department of Chemistry,
Tulane University,
New Orleans,
Louisiana 70188,
U.S.A.
e-mail: mlevy@tulane.edu

E.V. Ludeña,
Centro de Química Instituto Venezolano de Investigaciones Científicas,
IVIC,
Apartado 21827,
Caracas 1020-A,
VENEZUELA

R. Magyar,
Department of Physics and Astronomy,
Rutgers University,
136 Frelinghuysen Road,
Piscataway, NJ 08854-8019,
U.S.A.
e-mail: rmagyar@physics.rutgers.edu

R. McWeeny,
Dipartimento di Chimica e Chimica Industriale,
University of Pisa,
Via Risorgimento 35,
56100 Pisa,
ITALY
e-mail: mcweeny@dcci.unipi.it

C. Morrison,
Department of Chemistry,
University of Edinburgh,
Edinburgh,
SCOTLAND
e-mail: C.Morrison@ed.ac.uk

Á. Nagy,
Department of Theoretical Physics,
University of Debrecen,
H-4010 Debrecen,
HUNGARY
e-mail: anagy@madget.atomki.hu

E. Pérez-Romero,
Facultad de Quimica,
Universidad de Salamanca,
37008 Salamanca,
SPAIN

K. Refson,
Rutherford Appleton Laboratory,
Chilton,
Didcot,
Oxfordshire,
OX11 0QX,
ENGLAND
e-mail: K. Refson@rl.ac.uk

P. Rushton,
University of Durham,
Science Laboratories,
South Road,
Durham DH1 3LE,
ENGLAND

M. Seidl,
University of Regenbergs,
D93040 Regenbergs,
GERMANY

B.T. Sutcliffe,
Laboratoire de Chimie Physique Moléculaire,
Université Libre de Bruxelles,
B-1050 Bruxelles,
BELGIUM
e-mail: bsutclif@ulb.ac.be

L.M. Tel,
Departamento de Química Física,
Universidad de Salamanca,
37008 Salamanca, SPAIN

A.K. Theophilou,
DEMOCRITOS National Center for Scientific Research,
Institute of Materials Science,
TT 15310,
Athens,
GREECE
e-mail: theo@ims.demokritos.gr

D.J. Tozer,
Department of Chemistry,
University of Durham,
Durham,
ENGLAND
e-mail: D.J.Tozer@Durham.ac.uk

P. Tulip

R. van Leeuwen,
Theoretical Chemistry,
Materials Science Centre,
Rijksuniversiteit Groningen,
Nijenborgh 4 9747 AG,
Groningen,
THE NETHERLANDS

C.C. Wilson,
ISIS,
Rutherford Appleton Laboratory,
Chilton,
Didcot,
Oxfordshire,
OX11 0QX,
ENGLAND
e-mail: C.C.Wilson@rl.ac.uk

S. Wilson,
Atlas Centre,
Rutherford Appleton Laboratory,
Chilton,
Didcot,
Oxfordshire,
OX11 0QX,
ENGLAND
e-mail: S.Wilson@rl.ac.uk

II

THE PROCEEDINGS

THE KELDYSH FORMALISM APPLIED TO TIME-DEPENDENT CURRENT-DENSITY-FUNCTIONAL THEORY

Robert van Leeuwen
Theoretical Chemistry, Materials Science Centre
Rijksuniversiteit Groningen
Nijenborgh 4, 9747 AG, Groningen, The Netherlands

Abstract   In this work we demonstrate how to derive the Kohn-Sham equations of time-dependent current-density functional theory from a generating action functional defined on a Keldysh time contour. These Kohn-Sham equations contain an exchange-correlation contribution to the vector potential. For this quantity we derive an integral equation. We further derive an integral equation for its functional derivative, the exchange-correlation kernel, which plays an essential role in response theory. The exchange-only limits of the latter equation is studied in detail for the electron gas and future applications are discussed.

1.   Introduction

Time-dependent density-functional theory (TDDFT) [1,2] is a method for calculating properties of many-electron systems in time-dependent external fields. The theory has originally been formulated for systems in longitudinal electric fields, which can be described by a time-dependent scalar potential $v(\mathbf{r}t)$ (for reviews see [3–5]). The basis of this theory is the Runge-Gross theorem [2] which states that, for a given initial state, the external potential $v(\mathbf{r}t)$ (modulo an arbitrary purely time-dependent function $C(t)$) is a functional of the time-dependent density $n(\mathbf{r}t)$. This implies that every observable is a functional of the density and the initial state. The next step in TDDFT is then to introduce a noninteracting system with the same density $n(\mathbf{r}t)$ as the true system. This system is called the Kohn-Sham system and its external potential, which is a functional of the density, is denoted as $v_s(\mathbf{r}t)$. If one assumes that the initial state exists, then an explicit construction procedure for $v_s(\mathbf{r}t)$ can be given [5–7]. Once an approximation for $v_s(\mathbf{r}t)$ as a functional of the density is given, the Kohn-Sham equations can be solved self-consistently. This time-dependent Kohn-Sham theory has had many successful applications and is widely used.

Nevertheless, the theory has some limitations and drawbacks. First of all, the theory only applies to systems for which the external field can be described by a scalar potential, i.e. longitudinal fields. This excludes, for instance, the important case of external light fields, at least if we want to go beyond the dipole approximation. Secondly, the theory can not deal with infinite systems with periodic boundary conditions as one does in standard band structure methods [8]. The reason is that we try to describe the properties of a finite sample of material (macroscopic on a atomic scale) with the idealized infinite system. As a consequence the long range effects which result from the boundary charges must be externally reintroduced in a theory in which the basic variable is the ground state density of the bulk. This therefore requires an extension of TDDFT. Thirdly, it was shown by Vignale and Kohn [9, 10] that if one wants to go beyond the simple adiabatic local density approximation and include memory effects into the density functionals, one consequently needs functionals that are very nonlocal in space. Although this is not a problem in principle, it can be a practical problem for constructing reliable approximations. A theory that does not suffer from these three drawbacks is time-dependent current-density-functional theory (TDCDFT). In this theory the time-dependent current $\mathbf{j}(\mathbf{r}t)$, rather than the density, is the basic variable. In this theory the pair of scalar and vector potentials $(v(\mathbf{r}t), \mathbf{A}(\mathbf{r}t))$ rather than the scalar potential $v(\mathbf{r}t)$ alone plays a crucial role. As in the case of TDDFT we can derive a theorem of the Runge-Gross type [4, 11]. For a given initial state the time-dependent current density $\mathbf{j}(\mathbf{r}t)$ uniquely determines the applied potentials $(v(\mathbf{r}t), \mathbf{A}(\mathbf{r}t))$ up to a gauge transformation. This implies that all physical observables are functionals of the current density and the initial state. On the basis of this theorem we can guarantee the uniqueness of a Kohn-Sham system. This Kohn-Sham system is defined to be that system which has the same current as the true system. On the basis of the continuity equation

$$\partial_t n(\mathbf{r}t) + \nabla \cdot \mathbf{j}(\mathbf{r}t) = 0 \qquad (1)$$

and knowledge of the initial state it will also have the same density as the true system. The Kohn-Sham system will now have an external pair of potentials $(v_s(\mathbf{r}t), \mathbf{A}_s(\mathbf{r}t))$ which are functionals of the current-density. Once an approximation of these potentials as a functional of the current-density is given the Kohn-Sham equations can be solved self-consistently, and the current-density can be calculated. Such an ap-

The Keldysh formalism 45

proach has several advantages. First of all, the theory can be applied to transverse perturbing fields described by an vector potential $\mathbf{A}(\mathbf{r}t)$ and therefore presents an extension of the TDDFT formulation which only dealt with longitudinal fields. Secondly, the theory can be applied to calculate response properties of infinite systems, in which the changes of the boundary charges are measured by the current and from which the polarizability can be directly calculated [12–15] Thirdly, it has been shown by Vignale and Kohn [9] that TDCDFT allows for functionals that are nonlocal in time, but local in space. This is related to the fact that a gradient expansion exists for the current-current response function, whereas it does not for the density-density response function. An approximate current functional derived by Vignale and Kohn was shown to be very successful for the calculation of plasmon linewidths in quantum wells [16,17] and the polarizabilities of polymers [18].

The aim of this paper is to provide a systematic method to calculate current functionals to be used in the time-dependent Kohn-Sham equations. We will also fill in some details on the Keldysh formalism that were left out in earlier work [5,19,20]. The paper is divided as follows. We first introduce the Keldysh action functional and show how this functional can be used to generate the current-current response functions. We then show how the Keldysh functional can be used to derive the Kohn-Sham and corresponding linear response equations. By expanding this functional in terms of the two-particle interaction we obtain a diagrammatic perturbation expansion in terms of Keldysh Green functions for the effective vector potential and its functional derivative, to be used in the Kohn-Sham equations or their linearized form. The procedure is illustrated for the electron gas. We finally present an outlook and conclusions.

2. Keldysh action

We consider a system of interacting electrons in a time-dependent external field. A typical example would for instance be an atom or moleculein a laser field. The Hamiltonian which describes this system is given by

$$\hat{H}(t) = \hat{T} + \hat{W} + \hat{U}(t) \tag{2}$$

where the constituent term are given as follows. The kinetic energy $\hat{T}$ and the two-particle interaction $\hat{W}$ are given by

$$\hat{T} = -\sum_{\sigma} \tfrac{1}{2} \int d^3r\, \hat{\psi}_\sigma^\dagger(\mathbf{r}) \nabla^2 \hat{\psi}_\sigma(\mathbf{r}) \tag{3}$$

and
$$\hat{W} = \sum_{\sigma,\sigma'} \tfrac{1}{2} \int d^3r d^3r' \hat{\psi}^\dagger_\sigma(\mathbf{r}) \hat{\psi}^\dagger_{\sigma'}(\mathbf{r}') w(\mathbf{r},\mathbf{r}') \hat{\psi}_{\sigma'}(\mathbf{r}') \hat{\psi}_\sigma(\mathbf{r}) \qquad (4)$$

where w represent the usual Coulomb repulsion between the electrons. The term $\hat{U}(t)$ represents the time-dependent external field

$$\hat{U}(t) = \int d^3r (\hat{n}(\mathbf{r}) v_0(\mathbf{r}) + \hat{\mathbf{j}}_p(\mathbf{r}) \cdot \mathbf{A}(\mathbf{r}t) + \tfrac{1}{2}\hat{n}(\mathbf{r}) \mathbf{A}^2(\mathbf{r}t)) \qquad (5)$$

where

$$\hat{n}(\mathbf{r}) = \sum_\sigma \hat{\psi}^\dagger_\sigma(\mathbf{r}) \hat{\psi}_\sigma(\mathbf{r}) \qquad (6)$$

$$\hat{\mathbf{j}}_p(\mathbf{r}) = \sum_\sigma \frac{1}{2i} (\hat{\psi}^\dagger_\sigma(\mathbf{r}) \nabla \hat{\psi}_\sigma(\mathbf{r}) - [\nabla \hat{\psi}^\dagger_\sigma(\mathbf{r})] \hat{\psi}_\sigma(\mathbf{r})) \qquad (7)$$

are the density and the paramagnetic current operator. In the following, where we will mostly consider the linear response case, $v_0(\mathbf{r})$ will be the external potential in the ground state and all time-dependence will be incorporated in $\mathbf{A}(\mathbf{r}t)$. We have now completely defined our physical system. This leaves us with the task of calculating the physical quantities of interest.We are mostly interested in calculating the response of the system due toexternal perturbations. We therefore want to introduce a generating functionalfor the response functions. This approach is analogous to the situation instatistical mechanics where all thermodynamic quantities can be calculatedas derivatives with respect to parameters in the partition function.The basic idea is therefore to start with a generating functional resembling thepartition function of statistical mechanics.This functional should be constructed in such a way thatits first order derivative with respect to the external field yields the physical current density and its higher order derivatives yield the desired higher order response functions.

In earlier work we showed how such a functional can be constructed. This functional was defined on using the so-called Keldysh time contour. This time contour technique was introduced by Schwinger [21] and used by Keldysh [22] to obtain a diagrammatic perturbation expansion for nonequilibrium systems. In this technique the real time $t(\tau)$ is parametrized by an underlying pseudotime parameter τ [23–26]. The parametrization is such that, if the pseudotime time runs from a certain initial time τ_i to a final time τ_f, then the real time runs from t_0 to t_1 and back to t_0. This trick has the great virtue that, when evolution operators are used that are τ-ordered rather than t-ordered, then in perturbation theory Wick's theorem can be used, without invoking any adiabatic switch-off of the two-particle interactions as is done in the Gellman-Low theorem [27, 28]. It is exactly this feature that makes the theory applicable to nonstationary systems. Within time-dependent density functional theory it has already been shown that the formalism can also be used to resolve a paradox [5, 20] involving the symmetry and causality properties of response functions. Here we make an extension of that work to the case of current functionals.

We define a functional of the external field $\mathbf{A}$ by

$$\tilde{F}[\mathbf{A}] = i \ln \langle \Psi_0 | V(\tau_f, \tau_i) | \Psi_0 \rangle \quad (8)$$

where V is the τ- or contour ordered evolution operator of the system

$$V(\tau_2, \tau_1) = T_C \exp\left[-i \int_{\tau_1}^{\tau_2} d\tau t'(\tau) \hat{H}(\tau)\right] \quad (9)$$

and where T_C denotes ordering in τ and the state Ψ_0 is the initial state. Here we extend the definition of the Hamiltonian and allow $\mathbf{A}(\mathbf{r}\tau)$ to take different values of the forward and backward parts of the contour. It is clear from this equation that if the external field $\mathbf{A}$ is equal on the forward and backward parts of the contour, i.e. depending on τ only through $t(\tau)$, then this evolution will become unity and $\tilde{F}$ will become zero. Vector potentials of this type will in the following be denoted as physical vector potentials. The functional derivatives with respect to $\mathbf{A}$ taken at a physical vector potential will, however, be nonzero in general. When taking functional derivatives we use the basic equation

$$\frac{\delta}{\delta \mathbf{A}(\mathbf{r}\tau)} V(\tau_2, \tau_1) = -i V(\tau_2, \tau)(\hat{\mathbf{j}}_p(\mathbf{r}) + \hat{n}(\mathbf{r})\mathbf{A}(\mathbf{r}\tau)) V(\tau, \tau_1) \quad (10)$$

where $\tau_1 < \tau < \tau_2$. These equations follow directly from the Schrödinger equation. Using these equations we find

$$\frac{\delta \tilde{F}}{\delta \mathbf{A}(\mathbf{r}\tau)} = \frac{\langle \Psi_0 | V(\tau_f, \tau)(\hat{\mathbf{j}}_p(\mathbf{r}) + \hat{n}(\mathbf{r})\mathbf{A}(\mathbf{r}\tau)) V(\tau, \tau_i) | \Psi_0 \rangle}{\langle \Psi_0 | V(\tau_f, \tau_i) | \Psi_0 \rangle}$$

$$= \langle \hat{\mathbf{j}}_{p,H}(\mathbf{r}\tau) \rangle + \langle \hat{n}_H(\mathbf{r}\tau) \rangle \mathbf{A}(\mathbf{r}\tau) = \mathbf{j}(\mathbf{r}\tau) \quad (11)$$

where we defined the Heisenberg representation of an operator as usual by $\hat{O}_H(\tau) = V(\tau_i, \tau) \hat{O} V(\tau, \tau_i)$ and the expectation value by

$$\langle \hat{O}_H(\tau) \rangle = \frac{\langle \Psi_0 | T_C [V(\tau_f, \tau_i) \hat{O}_H(\tau)] | \Psi_0 \rangle}{\langle \Psi_0 | V(\tau_f, \tau_i) | \Psi_0 \rangle} \quad (12)$$

If we evaluate the derivative $\delta \tilde{F}/\delta \mathbf{A}$ at a physical potential $\mathbf{A}(\mathbf{r}t(\tau))$ we obtain

$$\frac{\delta \tilde{F}}{\delta \mathbf{A}(\mathbf{r}\tau)}\Big|_{\mathbf{A}=\mathbf{A}(\mathbf{r}t_1)} = \langle \Psi_0 | U(t_0,t_1)(\hat{\mathbf{j}}_p(\mathbf{r}) + \hat{n}(\mathbf{r})\mathbf{A}(\mathbf{r}t_1))U(t_1,t_0)|\Psi_0\rangle$$
$$= \mathbf{j}(\mathbf{r}t_1) \qquad (13)$$

where the evolution operator in real time is defined as usual by

$$U(t_1,t_0) = T\exp\left[-i\int_{t_0}^{t_1} dt \hat{H}(t)\right] \qquad (14)$$

Therefore the derivative of $\tilde{F}$ at the physical potential $\mathbf{A}$ is the gauge invariant current $\mathbf{j}(\mathbf{r}t)$ of the system in the external field $\mathbf{A}(\mathbf{r}t)$. We can now calculate higher order response functions by repeated differentiation. At this point it will be convenient to introduce a shortened notation. We will write $\bar{k} = \mathbf{r}_k \tau_k$ and $k = \mathbf{r}_k t_k$ and will also drop the subindex H from the operators. The current-current response function $\chi_{\mu\nu}$ is then given by

$$\chi_{\mu\nu}(\bar{1},\bar{2}) = \frac{\delta^2 \tilde{F}}{\delta A_\mu(\bar{1})\delta A_\nu(\bar{2})}$$

$$= \delta_{\mu\nu} n(\bar{1})\delta_C(1-2) - i\langle T_C[\Delta\hat{\mathbf{j}}_p(\bar{1})\Delta\hat{\mathbf{j}}_p(\bar{2})]\rangle$$

$\delta_C(t_1-t_2) = \delta(\tau_1-\tau_2)/t'(\tau_1)$ is the contour delta function and where the current-density fluctuation operator $\Delta\hat{\mathbf{j}}_{p,\mu}(\bar{1}) = \hat{\mathbf{j}}_{p,\mu}(\bar{1}) - \langle \hat{\mathbf{j}}_{p,\mu}(\bar{1})\rangle$ enters due to the derivatives of the denominator in Eq.(11). The response function is a symmetric function of its arguments as it should (being a second order functional derivative) and can be regarded as a integral kernel in pseudotime. It will however become a retarded function acting in physical time. In order to see this we calculate the current response $\delta \mathbf{j}(\mathbf{r}t)$ due to a potential variation $\delta \mathbf{A}(\mathbf{r}t)$. The density response function $\chi_{\mu\nu}$ evaluated at a physical currentdensity $\mathbf{j}(\mathbf{r}t)$ is given by

$$\chi_{\mu\nu}(\bar{1},\bar{2}) = \delta_{\mu\nu} n(1)\delta_C(1-2)$$
$$- i\theta(\tau_1-\tau_2)\langle \Delta\hat{\mathbf{j}}_{p,\mu}(1)\Delta\hat{\mathbf{j}}_{p,\nu}(2)\rangle$$
$$- i\theta(\tau_2-\tau_1)\langle \Delta\hat{\mathbf{j}}_{p,\nu}(2)\Delta\hat{\mathbf{j}}_{p,\mu}(1)\rangle \qquad (16)$$

Hence we have

$$\begin{aligned}\delta j_\mu(\mathbf{r}_1 t_1) &= \sum_\nu \int_{\tau_i}^{\tau_f} d\tau_2 t'(\tau_2) d^3 r_2 \chi_{\mu\nu}(\mathbf{r}_1\tau_1,\mathbf{r}_2\tau_2)\delta A_\nu(\mathbf{r}_2 t_2)\\
&= n_0(\mathbf{r}_1)\delta A_\nu(\mathbf{r},t_1)\\
&\quad - \sum_\nu i \int_{\tau_i}^{\tau_1} d\tau_2 t'(\tau_2) d^3 r_2 \langle \Delta\hat{\mathbf{j}}_{p,\mu}(\mathbf{r}_1 t_1)\Delta\hat{\mathbf{j}}_{p,\nu}(\mathbf{r}_2 t_2)\rangle \delta A_\nu(\mathbf{r}_2 t_2)\\
&\quad - \sum_\nu i \int_{\tau_1}^{\tau_f} d\tau_2 t'(\tau_2) d^3 r_2 \langle \Delta\hat{\mathbf{j}}_{p,\nu}(\mathbf{r}_2 t_2)\Delta\hat{\mathbf{j}}_{p,\mu}(\mathbf{r}_1 t_1)\rangle \delta A_\nu(\mathbf{r}_2 t_2)\\
&= \sum_\nu \int_{t_0}^{+\infty} dt_2 d^3 r_2 \chi_{R,\mu\nu}(\mathbf{r}_1 t_1,\mathbf{r}_2 t_2)\delta A_\nu(\mathbf{r}_2 t_2)\end{aligned}\qquad(17)$$

where

$$\begin{aligned}\chi_{R,\mu\nu}(1,2) &= n_0(\mathbf{r}_1)\delta_{\mu\nu}\delta(t_1-t_2)\delta(\mathbf{r}_1-\mathbf{r}_2)\\
&\quad - i\theta(t_1-t_2)\langle\Psi_0|[\hat{\mathbf{j}}_{p,\mu}(1),\hat{\mathbf{j}}_{p,\nu}(2)]|\Psi_0\rangle\end{aligned}\qquad(18)$$

In the last step we used that the expectation value of the commutator of the current fluctuation operators is equal to the expectation value of the commutator of the current operators themselves. The function $\chi_{R,\mu\nu}$ is the usual retarded response function as it usually appears in response theory. The outlined procedure applies to all higher order derivatives as well, i.e. all higher order response functions are symmetric functions in pseudotime and become causal or retarded functions in physical time (see Appendix D for an explicit example).

3. Kohn-Sham equations and linear response

We now want to use the current $\mathbf{j}(\mathbf{r}\tau)$ as our basic variable and we perform a Legendre transform and define

$$F[\mathbf{j}] = -\tilde{F}[\mathbf{A}] + \int_C dt d^3 r \mathbf{j}(\mathbf{r}\tau)\cdot\mathbf{A}(\mathbf{r}\tau)\qquad(19)$$

so that

$$\delta F/\delta \mathbf{j}(\mathbf{r}\tau) = \mathbf{A}(\mathbf{r}\tau)\qquad(20)$$

For convenience we introduced the short notation $\int_C dt$ for $\int d\tau t'(\tau)$. The Legendre transformation assumes that there is a one-to-one relation between $\mathbf{j}(\mathbf{r}\tau)$ and $\mathbf{A}(\mathbf{r}\tau)$ such that Eq. (11) is invertible (up to gauge). One can prove, when considering a system initially in its ground state, that the Keldysh current-current reponse function is invertible for switch-on processes. This can be done by a direct generalization of the proof given for the density-density response function in ref. [5]. We now define an action functional for a noninteracting system with the Hamiltonian

$$\hat{H}_s(\tau) = \hat{T} + \hat{U}_s(\tau) \tag{21}$$

where U_s has a form analogous to Eq.(5), and the action

$$\tilde{F}_s[\mathbf{A}_s] = i \ln \langle \Phi_0 | V_s(\tau_f, \tau_i) | \Phi_0 \rangle \tag{22}$$

The evolution operator $V_s(\tau_f, \tau_i)$ is defined similarly as in Eq.(9) with $\hat{H}$ replaced by $\hat{H}_s$. The initial wave function Φ_0 at $t = t_0$ is a noninteracting state and will often be a Slater determinant. We can now do a similar Legendre transform and define

$$F_s[\mathbf{j}] = -\tilde{F}_s[\mathbf{A}_s] + \int_C dt d^3 r \mathbf{j}(\mathbf{r}\tau) \cdot \mathbf{A}_s(\mathbf{r}\tau) \tag{23}$$

The exchange-correlation part F_{xc} of the action functional is then defined by

$$F[\mathbf{j}] = F_s[\mathbf{j}] - F_{xc}[\mathbf{j}] - \frac{1}{2} \int_C dt d^3 r_1 d^3 r_2 \frac{n(\mathbf{r}_1 \tau) n(\mathbf{r}_2 \tau)}{|\mathbf{r}_1 - \mathbf{r}_2|} \tag{24}$$

where the Keldysh density $n(\mathbf{r}\tau)$ is a functional of the initial state and the current $\mathbf{j}$ through the continuity equation

$$\partial_t n(\mathbf{r}\tau) + \nabla \cdot \mathbf{j}(\mathbf{r}\tau) = 0 \tag{25}$$

(note that $\partial_t = 1/t'(\tau) \partial_\tau$) We implicitly assume that the functionals F and F_s are defined on the same domain, i.e., that there exists a noninteracting system described by the Hamiltonian $\hat{H}_s$ with the same current density as the interacting system described by the Hamiltonian $\hat{H}$. A necessary requirement in order for this to be true is that the initial states Ψ_0 and Φ_0 must yield the same current density. For most applications, Ψ_0 will be the ground state of the system before the time-dependent field is switched on and Φ_0 will be the corresponding Kohn-Sham determinant of stationary density-functional theory. Functional differentiation of Eq. (24) with respect to $\mathbf{j}(\mathbf{r}\tau)$ yields

$$\mathbf{A}(\mathbf{r}\tau) = \mathbf{A}_s(\mathbf{r}\tau) - \mathbf{A}_{xc}(\mathbf{r}\tau) - \mathbf{A}_H(\mathbf{r}\tau) \tag{26}$$

where

$$\partial_t \mathbf{A}_H(\mathbf{r}\tau) = -\nabla \int d^3 r' \frac{n(\mathbf{r}'\tau)}{|\mathbf{r} - \mathbf{r}'|} \tag{27}$$

is the Hartree part of the vector potential and

$$\mathbf{A}_{xc}(\mathbf{r}\tau) = \delta F_{xc}/\delta \mathbf{j}(\mathbf{r}\tau) \tag{28}$$

is the exchange-correlation potential. By construction the vector potential $\mathbf{A}_s$ of the noninteracting system yields the same density as the vector potential $\mathbf{A}$ in the fully interacting system. The noninteracting system is thus to be identified with the time-dependent Kohn-Sham system. If we take the functional derivatives at the physical time-dependent current density $\mathbf{j}(\mathbf{r}t)$ corresponding to the vector potential $\mathbf{A}(\mathbf{r}\tau) = \mathbf{A}(\mathbf{r}t(\tau))$ of the interacting system, we can transform to physical time and the Kohn-Sham system is then given by the equations

$$(-\frac{1}{2}(\nabla + i(\mathbf{A} + \mathbf{A}_H + \mathbf{A}_{xc}))^2 + v_0(\mathbf{r}))\phi_i(\mathbf{r}t) = i\partial_t \phi_i(\mathbf{r}t)$$

$$\mathbf{A}_{xc}(\mathbf{r}t) = \frac{\delta F_{xc}}{\delta \mathbf{j}(\mathbf{r}\tau)}\Big|_{\mathbf{j}=\mathbf{j}(\mathbf{r}t)} \qquad (29)$$

where the density $\mathbf{j}(\mathbf{r}t)$ can be calculated from the orbitals according to

$$\mathbf{j}(\mathbf{r}t) = n(\mathbf{r}t)\mathbf{A}_s(\mathbf{r}t) + \frac{1}{2i}\sum_i^N (\phi_i^*(\mathbf{r}t)\nabla\phi_i(\mathbf{r}t) - (\nabla\phi_i^*(\mathbf{r}t))\phi_i(\mathbf{r}t)) \qquad (30)$$

Let us now see how the current-current response function can be obtained from this formalism. With the chain rule for differentiation we can write

$$\frac{\delta j_\mu(\bar{1})}{\delta A_\nu(\bar{2})} = \sum_\lambda \int_C d3 \frac{\delta j_\mu(\bar{1})}{\delta A_{s,\lambda}(\bar{3})} \frac{\delta A_{s,\lambda}(\bar{3})}{\delta A_\nu(\bar{2})} = \frac{\delta j_\mu(\bar{1})}{\delta A_{s,\nu}(\bar{2})}$$
$$+ \sum_{\lambda\kappa} \int_C d3d4 \frac{\delta j_\mu(\bar{1})}{\delta A_{s,\lambda}(\bar{3})} \frac{\delta(A_{H,\lambda}(\bar{3}) + A_{xc,\lambda}(\bar{3}))}{\delta j_\kappa(\bar{4})} \frac{\delta j_\kappa(\bar{4})}{\delta A_\nu(\bar{2})} \qquad (31)$$

Hence we obtain

$$\chi_{\mu\nu}(\bar{1},\bar{2}) = \chi_{s,\mu\nu}(\bar{1},\bar{2})$$
$$+ \sum_{\lambda\kappa} \int_C d3d4 \chi_{s,\mu\lambda}(\bar{1},\bar{3}) f_{Hxc,\lambda\kappa}(\bar{3},\bar{4}) \chi_{\kappa\nu}(\bar{4},\bar{2}) \qquad (32)$$

where we defined

$$f_{H,\mu\nu}(\bar{1},\bar{2}) = \frac{\delta A_{H,\mu}(\bar{1})}{\delta j_\nu(\bar{2})} \qquad (33)$$

$$f_{xc,\mu\nu}(\bar{1},\bar{2}) = \frac{\delta A_{xc,\mu}(\bar{1})}{\delta j_\nu(\bar{2})} \qquad (34)$$

$$f_{Hxc,\mu\nu}(\bar{1},\bar{2}) = f_{H,\mu\nu}(\bar{1},\bar{2}) + f_{xc,\mu\nu}(\bar{1},\bar{2}) \qquad (35)$$

We have therefore found a relation between the full and the Kohn-Sham current-current response function. A similar equation has been derived before within the context of pure density functional theory [29]. Equation (32) can now be transformed to physical time (for details see Appendix C) giving

$$\begin{aligned}\chi_{R,\mu\nu}(1,2) &= \chi_{R,s,\mu\nu}(1,2) \\ &+ \sum_{\lambda\kappa}\int d3d4\chi_{R,s,\mu\lambda}(1,3)f_{R,Hxc,\lambda\kappa}(3,4)\chi_{R,\kappa\nu}(4,2)\end{aligned} \quad (36)$$

where the Keldysh response functions are now replaced with the retarded causal ones. If we consider the response of a system initially in its ground state then the response function depends on the difference of the time coordinates. In that case it is convenient to Fourier transform to frequency space and we can write

$$\chi_{R,\mu\nu}(\mathbf{r}_1,\mathbf{r}_2,\omega) = \chi_{R,s,\mu\nu}(\mathbf{r}_1,\mathbf{r}_2,\omega) + \sum_{\lambda\kappa}\int d^3r_3 d^3r_4 \chi_{R,s,\mu\lambda}(\mathbf{r}_1,\mathbf{r}_3,\omega)$$
$$\times (f_{R,xc,\lambda\kappa}(\mathbf{r}_3,\mathbf{r}_4,\omega) - \frac{1}{\omega^2}\partial_\lambda^3 \frac{1}{|\mathbf{r}_3-\mathbf{r}_4|}\partial_\kappa^4)\chi_{R,\kappa\nu}(\mathbf{r}_4,\mathbf{r}_2,\omega) \quad (37)$$

Here we used the continuity equation in

$$\partial_t^2 \delta A_H(\mathbf{r}\tau) = \nabla \int d^3r' \frac{\nabla' \cdot \delta\mathbf{j}(\mathbf{r}'\tau)}{|\mathbf{r}-\mathbf{r}'|} \quad (38)$$

to calculate $f_{R,H}(\mathbf{r}_3,\mathbf{r}_4,\omega)$ explicitly in frequency space. Eq.(37) is the basic equation of time-dependent current response theory. It can, in fact, be used to defined $f_{xc,\mu\nu}$, as has been done by Vignale and Kohn [9,10]. This equation, together with the time-dependent Kohn-Sham equations, are the main results of this section. The next step will be to obtain approximations for $\mathbf{A}_{xc}$ and for the xc-kernel $f_{R,xc,\mu\nu}$. How to obtain such a approximations will be the subject of the following sections.

4. TDOPM equations

We will now use the new formalism to derive the current extension of the time-dependent optimized potential method (TDOPM) [30].
The exchange-correlation part F_{xc} of the action functional can be expanded in terms of Keldysh Green functions [19] where the perturbing Hamiltonianis given by $\hat{H} - \hat{H}_s$. The Keldysh Green function is defined as

$$\begin{aligned}G_{\sigma\sigma'}(\bar{1},\bar{2}) &= -i\langle\Psi_0|T_C\hat{\psi}_{\sigma,H}(\bar{1})\hat{\psi}_{\sigma'}^\dagger(\bar{2})|\Psi_0\rangle \\ &= \theta(\tau_1-\tau_2)G_{\sigma,\sigma'}^>(\bar{1},\bar{2}) + \theta(\tau_2-\tau_1)G_{\sigma,\sigma'}^<(\bar{1},\bar{2})\end{aligned} \quad (39)$$

where we defined

$$G^>_{\sigma\sigma'}(\bar{1},\bar{2}) = -i\langle\Psi_0|\hat{\psi}_{\sigma,H}(\bar{1})\hat{\psi}^\dagger_{\sigma',H}(\bar{2})|\Psi_0\rangle \qquad (40)$$

$$G^<_{\sigma\sigma'}(\bar{1},\bar{2}) = i\langle\Psi_0|\hat{\psi}^\dagger_{\sigma',H}(\bar{2})\hat{\psi}_{\sigma,H}(\bar{1})|\Psi_0\rangle \qquad (41)$$

The expansion of the logarithm of the evolution operator yields the set of closed connected diagrams. Perturbation theory also requires an adiabatic switching-on of $\hat{H} - \hat{H}_s$ in the physical time interval $(-\infty, t_0)$ in order to connect the states Ψ_0 and Φ_0. This is however readily achieved by extending the Keldysh contour to $-\infty$ [19]. The expansion of the action is then given in terms of the Kohn-Sham Green function

$$G_s(\bar{1},\bar{2}) = \theta(\tau_1 - \tau_2)G^>_s(\bar{1},\bar{2}) + \theta(\tau_2 - \tau_1)G^<_s(\bar{1},\bar{2}) \qquad (42)$$

where

$$G^<_s(\bar{1},\bar{2}) = i\sum_{i=1}^{N/2} \phi^*_i(\bar{1})\phi_i(\bar{2}) \qquad (43)$$

$$G^>_s(\bar{1},\bar{2}) = -i\sum_{i>N/2} \phi^*_i(\bar{1})\phi_i(\bar{2}) \qquad (44)$$

where we consider the spin restricted case. If we restrict ourselves to the first order terms we find that the Hartree term and the term with $\mathbf{A} - \mathbf{A}_s$ cancel and obtain the exchange-only expression in terms of the Kohn-Sham Green function

$$\begin{aligned} F_x[\mathbf{j}] &= \int_C d1 \int_C d2 G_s(\bar{1},\bar{2}) v(1,2) G_s(\bar{2},\bar{1}) \\ &= \int_C dt d^3r_1 d^3r_2 \frac{G^<_s(\mathbf{r}_1\tau, \mathbf{r}_2\tau) G^<_s(\mathbf{r}_2\tau, \mathbf{r}_1\tau)}{|\mathbf{r}_1 - \mathbf{r}_2|} \\ &= -\sum_{ij}^{N/2} \int_C dt d^3r_1 d^3r_2 \frac{\phi^*_i(\mathbf{r}_1\tau)\phi_i(\mathbf{r}_2\tau)\phi_j(\mathbf{r}_1\tau)\phi^*_j(\mathbf{r}_2\tau)}{|\mathbf{r}_1 - \mathbf{r}_2|} \end{aligned} \qquad (45)$$

where $v(1,2) = \delta_C(t_1 - t_2)/|\mathbf{r}_1 - \mathbf{r}_2|$ and we use spin integrated Green functions. One sees that this functional is an implicit functional of $n(\mathbf{r}\tau)$ but an explicit functional of the orbitals. Going to higher order in $\hat{H} - \hat{H}_s$, the Keldysh perturbation expansion, in a similar way, leads to orbital dependent expressions for the correlation part F_c of the action (although there will also be implicit orbital dependence since higher order diagrams also contain terms that involve $\mathbf{A}_{xc}$). In the general case one may obtain $\mathbf{A}_{xc}$ from

$$A_{xc,\mu}(\bar{2}) = \int_C dt d^3r_1 \frac{\delta F_{xc}}{\delta A_{s,\nu}(\bar{1})} \frac{\delta A_{s,\nu}(\bar{1})}{\delta j_\mu(\bar{2})} \qquad (46)$$

Matrix multiplication by $\chi_{s,\mu\nu}$ and using the chain rule for differentiation yields (from now on we drop subindex s from the Green functions)

$$\sum_\nu \int_C dt_2 d^3r_2 \chi_{s,\mu\nu}(\bar{1},\bar{2}) A_{xc,\nu}(\bar{2})$$
$$= \int_C d2 \int_C d3 \frac{\delta F_{xc}}{\delta G(\bar{2},\bar{3})} \frac{\delta G(\bar{2},\bar{3})}{\delta A_{s,\mu}(\bar{1})}$$
$$= \int_C d2 \int_C d3 \frac{\delta F_{xc}}{\delta G(\bar{2},\bar{3})} G(\bar{2},\bar{1}) \mathbf{j}_{op,\mu}(\mathbf{r}_1) G(\bar{1},\bar{3}) \qquad (47)$$

where we defined the one-particle current operator $\mathbf{j}_{op,\nu} = 1/2i(\overrightarrow{\partial}_\mu - \overleftarrow{\partial}_\mu)$, where the second derivarive $\overleftarrow{\partial}_\mu)$ acts on the function to the left of the operator. We further used the relation (see Appendix B)

$$\frac{\delta G(\bar{1},\bar{2})}{\delta A_\nu(\bar{3})} = G(\bar{1},\bar{3}) \mathbf{j}_{op,\nu}(\mathbf{r}_3) G(\bar{3},\bar{2}) \qquad (48)$$

Let us now restrict ourselves to the exchange-only case. Then we obtain the integral equation

$$\sum_\nu \int_C dt_2 d^3r_2 \chi_{s,\mu\nu}(\bar{1},\bar{2}) A_{x,\nu}(\bar{2})$$
$$= 2 \int_C d2 \int_C d3 v(3,2) G(\bar{3},\bar{2}) G(\bar{2},\bar{1}) \mathbf{j}_{op,\mu}(\mathbf{r}_1) G(\bar{1},\bar{3}) \qquad (49)$$

An explicit expression for the Kohn-Sham response function in terms of the orbitals is given in Appendix A. If we transform to physical time we obtain the following equation for the x-only vector potential

$$\sum_\nu \int dt_2 d^3r_2 \chi_{R,s,\mu\nu}(1,2) A_{x,\nu}(2) = \Pi_\mu(1) \qquad (50)$$

where

$$\Pi_\mu(1) = 2 \int d^3r_2 \int d^3r_2 \int_{-\infty}^{t_1} dt \frac{G^<(\mathbf{r}_2 t, \mathbf{r}_3 t)}{|\mathbf{r}_2 - \mathbf{r}_3|}$$
$$\times \left[G^<(\mathbf{r}_2 t, \mathbf{r}_1 t) \mathbf{j}_{op,\mu}(\mathbf{r}_1) G^>(\mathbf{r}_1 t, \mathbf{r}_3 t) - (> \leftrightarrow <) \right] \qquad (51)$$

where the last term in brackets denotes that this term is obtain by interchanging $>$ and $<$ in the previous term. Note that the time integration in this expression runs to $-\infty$. This is due to the adiabatic connection between Kohn-Sham and true initial state before time t_0. The time-integration in the interval $[-\infty, t_0]$ can be done explicitly since the time-dependence of the Kohn-Sham orbitals is known in this time interval. For a further discussion of similar terms see ref. [31, 32]

Eqn.(50) together with (51) is the main result of this section. They are the current-density generalization of the well-known equations of the time-dependent optimized potential method (TDOPM). In case the external fields are longitudinal these equations can (using gauge transformation and the continuity equation) be shown to be equivalent to the TDOPM equations.

5. Integral equation for the xc-kernel

We will now study the integral equation for the xc-kernel $f_{xc,\mu\nu}$. Being a two-point function it gives much more information than just $\mathbf{A}_{xc}$. The determining equation for $f_{xc,\mu\nu}$ is consequently more complicated. It can be obtained by differentiation of Eq.(47) with respect to $A_{s,\kappa}$. This yields

$$\sum_\nu \int_C dt_2 d^3r_2 \frac{\delta \chi_{s,\mu\nu}(\bar{1},\bar{2})}{\delta A_{s,\kappa}(\bar{3})} A_{xc,\nu}(\bar{2})$$
$$+ \sum_\nu \int_C dt_2 d^3r_2 \chi_{s,\mu\nu}(\bar{1},\bar{2}) \frac{\delta A_{xc,\nu}(\bar{2})}{\delta A_{s,\kappa}(\bar{3})} = Q_{\mu\kappa}(\bar{1},\bar{3}) \qquad (52)$$

where we defined

$$Q_{\mu\kappa}(\bar{1},\bar{3}) = \frac{\delta^2 F_{xc}}{\delta A_{s,\kappa}(\bar{3}) \delta A_{s,\mu}(\bar{1})} \qquad (53)$$

With the chain rule for differentiation this can be written as

$$\sum_\nu \int_C dt_2 d^3r_2 \chi^{(2)}_{s,\mu\nu\kappa}(\bar{1},\bar{2},\bar{3}) A_{xc,\nu}(\bar{2})$$
$$+ \sum_{\nu\lambda} \int_C dt_2 d^3r_2 \int_C dt_4 d^3r_4 \chi_{s,\mu\nu}(\bar{1},\bar{2}) f_{xc,\nu\lambda}(\bar{2},\bar{4}) \chi_{s,\lambda\kappa}(\bar{4},\bar{3})$$
$$= Q_{\mu\kappa}(\bar{1},\bar{3}) \qquad (54)$$

where we defined the second order Keldysh response function of the Kohn-Sham system

$$\chi^{(2)}_{s,\mu\nu\kappa}(\bar{1},\bar{2},\bar{3}) = (-i)^2 \langle T_C \Delta j_{p,H,\mu}(\bar{1}) \Delta j_{p,H,\nu}(\bar{2}) \Delta j_{p,H,\kappa}(\bar{3}) \rangle \qquad (55)$$

Eq.(54) is an integral equation for the xc-kernel. In these equations the first and second order Kohn-Sham response functions $\chi_{s,\mu\nu}$ and $\chi^{(2)}_{s,\mu\nu\kappa}$ are explicitely known in terms of the Kohn-Sham orbitals (see Appendix

A for the explicit form of $\chi_{s,\mu\nu}$). The equation can now be transformed to physical time. The key steps to do this are explained in Appendix C and D. The result is

$$\sum_\nu \int dt_2 d^3r_2 \chi^{(2)}_{R,s,\mu\nu\kappa}(1,2,3) A_{xc,\nu}(2)$$

$$+ \sum_{\nu\lambda} \int dt_2 d^3r_2 \int dt_4 d^3r_4 \chi_{R,s,\mu\nu}(1,2) f_{R,xc,\nu\lambda}(2,4)\chi_{R,s,\lambda\kappa}(4,3)$$

$$= Q_{R,\mu\kappa}(1,3) \qquad (56)$$

where

$$\chi^{(2)}_{s,\mu\nu\kappa}(1,2,3) = (-i)^2 \theta(t_1-t_2)\theta(t_2-t_3)$$
$$\times \langle \Psi_0 |[[\hat{j}_{p,H,\mu}(1), \hat{j}_{p,H,\nu}(2)], \hat{j}_{p,H,\kappa}(3)]|\Psi_0\rangle \qquad (57)$$

is the retarded second order current response function. Eq.(56) can be solved self consistently once the $Q_{\mu\kappa}$ is given. In the following we will make the exchange-only approximation for this term, which amounts to an expansion to first order in the two-particle interaction. In that case the inhomogeneity $Q_{\mu\kappa}$ is given as

$$Q_{\mu\kappa}(\bar{1},\bar{3})$$
$$- \frac{\delta}{\delta A_{s,\kappa}(\bar{3})}\left(2\int_C d2 \int_C d4 v(2,4) G(\bar{2},\bar{4}) G(\bar{4},\bar{1})\mathbf{j}_{\mathbf{op},\mu}(\mathbf{r}_1) G(\bar{1},\bar{2})\right)$$
$$= 2\int_C d2 \int_C d4 G(\bar{2},\bar{3})\mathbf{j}_{\mathbf{op},\kappa}(\mathbf{r}_3) G(\bar{3},\bar{4})v(2,4) G(\bar{4},\bar{1})\mathbf{j}_{\mathbf{op},\mu}(\mathbf{r}_1) G(\bar{1},\bar{2})$$
$$+ 2\int_C d2 \int_C d4 G(\bar{2},\bar{4})v(2,4) G(\bar{4},\bar{1})\mathbf{j}_{\mathbf{op},\mu}(\mathbf{r}_1) G(\bar{1},\bar{3})\mathbf{j}_{\mathbf{op},\kappa}(\mathbf{r}_3) G(\bar{3},\bar{2})$$
$$+ 2\int_C d2 \int_C d4 G(\bar{2},\bar{4})v(2,4) G(\bar{4},\bar{3})\mathbf{j}_{\mathbf{op},\kappa}(\mathbf{r}_3) G(\bar{3},\bar{1})\mathbf{j}_{\mathbf{op},\mu}(\mathbf{r}_1) G(\bar{1},\bar{2})$$

Those three terms represent the three first order diagrams of the polarization bubble. They are presented graphically as Feynman diagrams in Figure 1. Each black line denotes a Green function and the wiggly line denotes the two-particle interaction. Let us now transform to physical time and work out the terms in $Q_{R,\mu\kappa}$ which consists of the three Feynman diagrams transformed to physical time. We write these three terms as

$$Q_{R,\mu\kappa} = D_{1,\mu\kappa} + D_{2,\mu\kappa} + D_{3,\mu\kappa} \qquad (58)$$

The Keldysh formalism

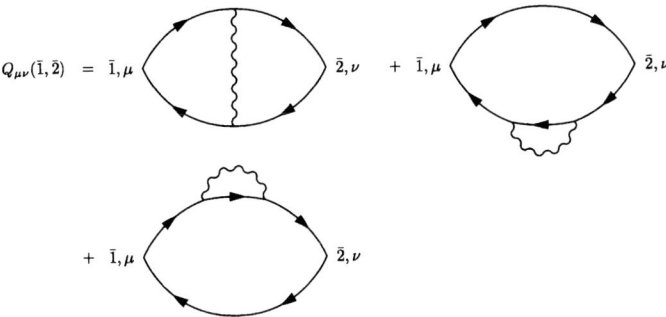

$Q_{\mu\nu}(\bar{1},\bar{2}) =$

Figure 1. The three first order polarization diagrams. The labels μ and ν denote insertion of the operator $j_{\mu,\text{op}}$ and $j_{\nu,\text{op}}$. The first diagram contains a vertex while the other two contain self-energy insertions.

The first diagram contains a vertex and yields in real time the expression

$$D_{1,\mu\kappa}(\mathbf{r}_1 t_1, \mathbf{r}_3 t_3) = \\ -2 \int_{t_0}^{\infty} dt \int d^3 r_2 d^3 r_4 \theta(t_1 - t)\theta(t - t_3) v(\mathbf{r}_2, \mathbf{r}_4) \\ \times [G^{>}(\mathbf{r}_4 t, \mathbf{r}_1 t_1)\mathbf{j}_{\text{op},\mu}(\mathbf{r}_1) G^{<}(\mathbf{r}_1 t_1, \mathbf{r}_2 t) - (> \leftrightarrow <)] \\ \times [G^{>}(\mathbf{r}_2 t, \mathbf{r}_3 t_3)\mathbf{j}_{\text{op},\kappa}(\mathbf{r}_3) G^{<}(\mathbf{r}_3 t_3, \mathbf{r}_4 t) - (> \leftrightarrow <)] \quad (59)$$

The remaining two diagrams contain self-energy insertions and are complex conjugates of each other, i.e. $D_{3,\mu\kappa} = D_{2,\mu\kappa}^*$, and therefore knowledge of one of them suffices to know the other. For $D_{3,\mu\kappa}$ we find the following explicit expression

$$D_{3,\mu\kappa} = A_{3,\mu\kappa} + B_{3,\mu\kappa} + C_{3,\mu\kappa} \quad (60)$$

where the three terms A, B and C are given by

$$A_{3,\mu\kappa}(\mathbf{r}_1 t_1, \mathbf{r}_3 t_3) = \\ 2 \int_{t_0}^{\infty} dt \int d^3 r_2 d^3 r_4 \theta(t_1 - t)\theta(t - t_3) v(\mathbf{r}_2, \mathbf{r}_4) G^{<}(\mathbf{r}_2 t, \mathbf{r}_4 t) \\ \times [G^{>}(\mathbf{r}_4 t, \mathbf{r}_3 t_3)\mathbf{j}_{\text{op},\kappa}(\mathbf{r}_3) G^{<}(\mathbf{r}_3 t_3, \mathbf{r}_1 t_1)\mathbf{j}_{\text{op},\mu}(\mathbf{r}_1) G^{>}(\mathbf{r}_1 t_1, \mathbf{r}_2 t) \\ + (> \leftrightarrow <)] \quad (61)$$

and

$$B_{3,\mu\kappa}(\mathbf{r}_1 t_1, \mathbf{r}_3 t_3) =$$
$$2 \int_{t_0}^{\infty} dt \int d^3 r_2 d^3 r_4 \theta(t_1 - t_3) \theta(t_3 - t) v(\mathbf{r}_2, \mathbf{r}_4) G^{<}(\mathbf{r}_2 t, \mathbf{r}_4 t)$$
$$\times \left[G^{>}(\mathbf{r}_4 t, \mathbf{r}_3 t_3) \mathbf{j}_{\text{op},\kappa}(\mathbf{r}_3) G^{>}(\mathbf{r}_3 t_3, \mathbf{r}_1 t_1) \mathbf{j}_{\text{op},\mu}(\mathbf{r}_1) G^{<}(\mathbf{r}_1 t_1, \mathbf{r}_2 t) \right.$$
$$\left. + (> \leftrightarrow <) \right] \tag{62}$$

and

$$C_{3,\mu\kappa}(\mathbf{r}_1 t_1, \mathbf{r}_3 t_3) =$$
$$-2 \int_{t_0}^{\infty} dt \int d^3 r_2 d^3 r_4 \theta(t_1 - t_3) \theta(t_1 - t) v(\mathbf{r}_2, \mathbf{r}_4) G^{<}(\mathbf{r}_2 t, \mathbf{r}_4 t)$$
$$\times \left[G^{>}(\mathbf{r}_4 t, \mathbf{r}_3 t_3) \mathbf{j}_{\text{op},\kappa}(\mathbf{r}_3) G^{<}(\mathbf{r}_3 t_3, \mathbf{r}_1 t_1) \mathbf{j}_{\text{op},\mu}(\mathbf{r}_1) G^{<}(\mathbf{r}_1 t_1, \mathbf{r}_2 t) \right.$$
$$\left. + (> \leftrightarrow <) \right] \tag{63}$$

With these explict equations the x-only kernel is completely defined. The explicit solution for the kernel for the case of the electron gas will discussed in the next section.

6. The exchange-only kernel for the electron gas

The general equations for the xc-kernel for arbitrary inhomogeneous systems is rather involved. The equations become, however, much more tractable for the homogeneous case for which they can be studied in more detail. Also the investigation of inhomogeneous systems benefits from such a study. The reason is that such kernels can in linear response theory be used in a local density approximation for the induced xc-vector potential $\delta \mathbf{A}_{xc}$ of the form

$$\delta A_{xc,\mu}(\mathbf{r},\omega) = \sum_\nu \int d^3 r f^h_{xc,R,\mu\nu}(n_0(\mathbf{r}); |\mathbf{r}-\mathbf{r}'|, \omega) \delta j_\nu(\mathbf{r}'\omega) \tag{64}$$

where we Fourier transformed from time to frequency space. Here $f^h_{xc,R,\mu\nu}$ is the xc-kernel of the homogeneous electron gas, n_0 is the ground state density and $\delta \mathbf{j}$ the induced current. In order to apply this local density approximation we need an explicit form for $f^h_{xc,R,\mu\nu}$. Since for the electron gas $f_{xc,\mu\nu}$ (from now on we will drop supindex h and subindex R, as we will only deal with the retarded electron gas kernel) depends on the difference between coordinates $\mathbf{r}$ and $\mathbf{r}'$ we can Fourier transform with respect to $\mathbf{r} - \mathbf{r}'$ and obtain the xc-kernel $f_{xc,\mu\nu}(\mathbf{q},\omega)$ in Fourier space. This function has been investigated by several researchers in the limit $|\mathbf{q}| \to 0$ for fixed finite ω (see [33] and references therein). In the present work we consider a different limit. We will study $f_{xc,\mu\nu}$ in the whole $q-\omega$ plane but restrict ourselves to the exchange-only limit. The corresponding function is denoted as $f_{x,\mu\nu}(\mathbf{q},\omega)$. From the previous section we see that it satisfies the following equation

$$\sum_{\nu\lambda} \chi_{R,s,\mu\nu}(\mathbf{q},\omega) f_{x,\nu\lambda}(\mathbf{q},\omega) \chi_{R,s,\lambda\kappa}(\mathbf{q},\omega) = Q_{\mu\kappa}(\mathbf{q},\omega) \tag{65}$$

The Kohn-Sham response function $\chi_{s,\mu\nu}$ and the true response function $\chi_{\mu\nu}$ have the following structure [34] (we also drop subindex R from response functions from now on)

$$\chi_{s,\mu\nu}(\mathbf{q},\omega) = \chi_{s,L}(\mathbf{q},\omega) \frac{q_\mu q_\nu}{q^2} + \chi_{s,T}(\mathbf{q},\omega)(\delta_{\mu\nu} - \frac{q_\mu q_\nu}{q^2}) \tag{66}$$

which defines the longitudinal and transverse response functions χ_L and χ_T. This structure follows from isotropy of the electron gas. The xc-kernel has a similar structure. We can therefore define longitudinal and transverse exchange response kernels by

$$f_{x,\mu\nu}(\mathbf{q},\omega) = \frac{1}{\omega^2} \left[f_{x,L}(\mathbf{q},\omega) q_\mu q_\nu + f_{x,T}(q,\omega)(q^2 \delta_{\mu\nu} - q_\mu q_\nu) \right] \tag{67}$$

If we insert this in Eq.(65) we obtain

$$\begin{aligned} Q_{\mu\kappa}(\mathbf{q},\omega) &= \frac{q_\mu q_\kappa}{q^2} \chi_{s,L}(\mathbf{q},\omega) \frac{q^2}{\omega^2} f_{x,L}(\mathbf{q},\omega) \chi_{s,L}(\mathbf{q},\omega) \\ &+ (\delta_{\mu\kappa} - \frac{q_\mu q_\kappa}{q^2}) \chi_{s,T}(\mathbf{q},\omega) \frac{q^2}{\omega^2} f_{x,T}(\mathbf{q},\omega) \chi_{s,T}(\mathbf{q},\omega) \end{aligned} \tag{68}$$

If we define longitudinal and transverse parts of $Q_{\mu\kappa}$ in the usual way, i.e.

$$Q_{\mu\nu}(\mathbf{q},\omega) = Q_L(\mathbf{q},\omega) \frac{q_\mu q_\nu}{q^2} + Q_T(\mathbf{q},\omega)(\delta_{\mu\nu} - \frac{q_\mu q_\nu}{q^2}) \tag{69}$$

then we obtain the equations

$$f_{x,L}(\mathbf{q},\omega) = \frac{\omega^2}{q^2} \frac{Q_L(\mathbf{q},\omega)}{\chi_{s,L}^2(\mathbf{q},\omega)} \tag{70}$$

$$f_{x,T}(\mathbf{q},\omega) = \frac{\omega^2}{q^2} \frac{Q_T(\mathbf{q},\omega)}{\chi_{s,T}^2(\mathbf{q},\omega)} \tag{71}$$

$$\tag{72}$$

We see that the equations for the longitudinal and transverse part completely decouple for the electron gas, as we would expect for an isotropic system. Since both $\chi_{s,L}$ and $\chi_{s,T}$ are known analytically (they were already calculated by Lindhard [35]) it remains to determine Q_L and Q_T. This can be done be working out the explicit formulas for the diagrams of the previous section. The longitudinal function Q_L has been worked out in detail before [36–38]. We therefore only present the expression for the transverse term:

$$\begin{aligned}Q_T(\mathbf{q},\omega) = & \int \frac{d^3k}{(2\pi)^3} \int \frac{d^3k'}{(2\pi)^3} v(\mathbf{k}-\mathbf{k}')(n_\mathbf{k}-n_{\mathbf{k}+\mathbf{q}})(n_{\mathbf{k}'}-n_{\mathbf{k}'+\mathbf{q}}) \\ & \times \left(\frac{(k^2-(\mathbf{k}\cdot\mathbf{q})^2/q^2)}{(\omega-\epsilon_{\mathbf{k}+\mathbf{q}}+\epsilon_\mathbf{k}+i\eta)^2} \right. \\ & \left. - \frac{(\mathbf{k}\cdot\mathbf{k}'-(\mathbf{k}\cdot\mathbf{q})(\mathbf{k}'\cdot\mathbf{q})/q^2)}{(\omega-\epsilon_{\mathbf{k}+\mathbf{q}}+\epsilon_\mathbf{k}+i\eta)(\omega-\epsilon_{\mathbf{k}'+\mathbf{q}}+\epsilon_{\mathbf{k}'}+i\eta)} \right)\end{aligned}$$

where $n_\mathbf{k} = \theta(k_F - |\mathbf{k}|)$ is the Fermi function in which k_F is the Fermi wave vector. The first term in brackets arises from the self-energy diagrams whereas the second term originates from the vertex diagram. The function Q_T can be reduced to a two-dimensional integral that can be evaluated numerically. Its properties are currently being investigated in detail [39], as well as its applicability in local density type approximations for inhomogeneous systems.

7. Conclusions

We showed how the basic equations of time-dependent current-density functional theory can be derived elegantly using the Keldysh formalism and provided some details that were left out in previous work. We also to derived an extension of the TDOPM equations that can be used in current-density-functional theory and found the corresponding equations for the xc-kernels, that play an essential role in response theory. We further exploited a key feature of the Keldysh formalism, namely the possibility to perform a systematic diagrammatic expansion for nonequilibrium systems. In this way we were able to derive an explicit expression for the transverse exchange kernel of the electron gas. This expression is currently being investigated in detail in order to obtain new functionals that can be used in calculations of response properties of inhomogeneous systems.

Acknowledgments

I like to thank Ulf von Barth and Carl-Olof Almbladh for many valuable discussions at the Solid State Theory Department of the University of Lund, where a large part of this work has been carried out.

Appendix: A

In this section we will calculate the response in the orbitals due to a small change in the vector potential of the Kohn-Sham equations in Keldysh form

$$\left(i\partial_t + \frac{1}{2}(\nabla + i\delta\mathbf{A}_s(\mathbf{r}\tau))^2 - v_0(\mathbf{r})\right)\varphi(\mathbf{r}\tau) = 0 \tag{A.1}$$

We write $\varphi = \phi + \delta\phi$ where ϕ_0 satisfies the Kohn-Sham equation without the external field $\delta\mathbf{A}_s$, i.e.

$$\left(i\partial_t + \frac{1}{2}\nabla^2 - v_0(\mathbf{r})\right)\phi(\mathbf{r}\tau) = 0 \tag{A.2}$$

With these equations we find that

$$\left(i\partial_t + \frac{1}{2}\nabla^2 - v_0(\mathbf{r})\right)\delta\phi(\mathbf{r}\tau) = \sum_\nu -\frac{i}{2}\partial_\nu(A_{s,\nu}\phi) - \frac{i}{2}A_{s,\nu}\partial_\nu\phi \tag{A.3}$$

where for the forward solution on the Keldysh contour we have initial value condition condition $\delta\phi(\mathbf{r}\tau_i) = 0$. This equation can be solved by expanding $\delta\phi$ in the unperturbed Kohn-Sham states $\phi_k(\mathbf{r}\tau) = \varphi_k(\mathbf{r})\exp(-i\epsilon_k t(\tau))$ where φ_k is an Kohn-Sham orbital of the unperturbed stationary system with eigenvalue ϵ_k. Thus we have

$$\delta\phi(\mathbf{r}\tau) = \sum_k c_k(\tau)\phi_k(\mathbf{r}\tau) \tag{A.4}$$

with initial value condition $c_k(\tau_i) = 0$. Substituting this into Eq.(A.3) and integrating over spatial coordinates after multiplying by $\phi_k^*(\mathbf{r}\tau)$ yields the solution

$$c_k(\tau_2) = -i\int_{\tau_i}^{\tau_2} d\tau_1 t'(\tau_1)\int d^3 r_1 \sum_\nu \frac{1}{2i}(\phi_k^*(\mathbf{r}_1\tau_1)\partial_\nu\phi(\mathbf{r}_1\tau_1) - \partial_\nu\phi_k^*(\mathbf{r}_1\tau_1)\phi(\mathbf{r}_1\tau_1))$$
$$\delta A_{s,\nu}(\mathbf{r}_1\tau_1) \tag{A.5}$$

We therefore find

$$\delta\phi(\mathbf{r}_2\tau_2) = -i\sum_k \int_{\tau_i}^{\tau_2} d\tau_1 t'(\tau_1)\int d^3 r_1 \sum_\nu [\phi_k^*(\mathbf{r}_1\tau_1)\mathbf{j}_{\mathrm{op},\nu}(\mathbf{r}_1)\phi(\mathbf{r}_1\tau_1)]\phi_k(\mathbf{r}_2\tau_2)$$
$$\delta A_{s,\nu}(\mathbf{r}_1\tau_1) \tag{A.6}$$

where we defined the one-particle current operator $\mathbf{j}_{\mathrm{op},\nu} = 1/2i(\overrightarrow{\partial}_\nu - \overleftarrow{\partial}_\nu)$, where the second derivarive $\overleftarrow{\partial}_\nu$) acts on the orbital tothe left of the operator. We thus conclude that

$$\frac{\delta\phi(\mathbf{r}_2\tau_2)}{\delta A_{s,\nu}(\mathbf{r}_1\tau_1)} = -i\theta(\tau_2 - \tau_1)\sum_k \phi_k(\mathbf{r}_2\tau_2)[\phi_k^*(\mathbf{r}_1\tau_1)\mathbf{j}_{\mathrm{op},\nu}(\mathbf{r}_1)\phi(\mathbf{r}_1\tau_1)] \tag{A.7}$$

We can carry out a similar procedure for the backward solutionon the Keldysh contour, which satisfies $\delta\phi(\mathbf{r}\tau_f) = 0$. With this boundary condition we find

$$\delta\phi^*(\mathbf{r}_2\tau_2) = -i\sum_k \int_{\tau_2}^{\tau_f} d\tau_1 t'(\tau_1)\int d^3 r_1 \sum_\nu [\phi_k^*(\mathbf{r}_1\tau_1)\mathbf{j}_{\mathrm{op},\nu}(\mathbf{r}_1)\phi(\mathbf{r}_1\tau_1)]^*\phi_k^*(\mathbf{r}_2\tau_2)$$
$$\delta A_{s,\nu}(\mathbf{r}_1\tau_1) \tag{A.8}$$

and therefore

$$\frac{\delta \phi^*(\mathbf{r}_2\tau_2)}{\delta A_{s,\nu}(\mathbf{r}_1\tau_1)} = -i\theta(\tau_1-\tau_2)\sum_k \phi_k^*(\mathbf{r}_2\tau_2)[\phi_k^*(\mathbf{r}_1\tau_1)\mathbf{j}_{op,\nu}(\mathbf{r}_1)\phi(\mathbf{r}_1\tau_1)]^*$$

$$= -i\theta(\tau_1-\tau_2)\sum_k \phi_k^*(\mathbf{r}_2\tau_2)[\phi^*(\mathbf{r}_1\tau_1)\mathbf{j}_{op,\nu}(\mathbf{r}_1)\phi_k(\mathbf{r}_1\tau_1)] \quad (A.9)$$

We have therefore obtained the desired equations for the orbital variations on the forward and backward parts of the Keldysh contour. These equations can now easily be used to calculate the Keldysh current response function of the Kohn-Sham system. Since

$$j_\mu(\mathbf{r}_1\tau_1) = n_0(\mathbf{r}_1)A_{s,\mu}(\mathbf{r}_1\tau_1) + \sum_{i=1}^N [\phi_i^*(\mathbf{r}_1\tau_1)\mathbf{j}_{op,\mu}(\mathbf{r}_1)\phi_i(\mathbf{r}_1\tau_1)] \quad (A.10)$$

we obtain using Eqns.(A.7) and (A.9)

$$\frac{\delta j_\mu(\mathbf{r}_1\tau_1)}{\delta A_{s,\nu}(\mathbf{r}_1\tau_1)} = \delta_{\mu\nu}n_0(\mathbf{r}_1)\delta(\mathbf{r}_1-\mathbf{r}_2)\delta_C(t_1-t_2)$$
$$+ \theta(\tau_1-\tau_2)\chi_{\mu\nu}^>(\mathbf{r}_1 t_1,\mathbf{r}_2 t_2) + \theta(\tau_2-\tau_1)\chi_{\mu\nu}^<(\mathbf{r}_1 t_1,\mathbf{r}_2 t_2) \quad (A.11)$$

where the functional derivative is evaluated at the physical vector potential $\mathbf{A}_s = 0$ and where we defined

$$\chi_{\mu\nu}^>(\mathbf{r}_1 t_1,\mathbf{r}_2 t_2) =$$
$$-i\sum_{i=1}^N \sum_k [\phi_i^*(\mathbf{r}_1 t_1)\mathbf{j}_{op,\mu}(\mathbf{r}_1)\phi_k(\mathbf{r}_1 t_1)][\phi_k^*(\mathbf{r}_2 t_2)\mathbf{j}_{op,\nu}(\mathbf{r}_2)\phi_i(\mathbf{r}_2 t_2)] \quad (A.12)$$

$$\chi_{\mu\nu}^<(\mathbf{r}_1 t_1,\mathbf{r}_2 t_2) = \chi_{\nu\mu}^>(\mathbf{r}_2 t_2,\mathbf{r}_1 t_1) \quad (A.13)$$

This defines the Kohn-Sham current response function in terms of the unperturbed orbitals.

Appendix: B

In this appendix we calculate the change in a Keldysh Green function due to a change in the vector potential. The zeroth order Green function satisfies

$$\left(i\partial_{t_1} + \frac{1}{2}\nabla_1^2 - v_0(\mathbf{r}_1)\right)G_0(\mathbf{r}_1\tau_1,\mathbf{r}_2\tau_2) = \delta_C(t_1-t_2) \quad (B.1)$$

We then want to solve

$$\left(i\partial_{t_1} + \frac{1}{2}(\nabla_1 + i\delta A_s(\mathbf{r}_1\tau_1))^2 - v_0(\mathbf{r}_1)\right)(G_0(\mathbf{r}_1\tau_1,\mathbf{r}_2\tau_2) + \delta G(\mathbf{r}_1\tau_1,\mathbf{r}_2\tau_2)) = \delta_C(t_1-t_2) \quad (B.2)$$

for δG. This gives, to first order in δG, the equation

$$\left(i\partial_{t_1} + \frac{1}{2}\nabla_1^2 - v_0(\mathbf{r}_1)\right)\delta G(\mathbf{r}_1\tau_1,\mathbf{r}_2\tau_2)$$
$$= -\frac{i}{2}\sum_\nu \partial_\nu^1(A_{s,\nu}(\mathbf{r}_1\tau_1)G_0(\mathbf{r}_1\tau_1,\mathbf{r}_2\tau_2)) + A_{s,\nu}(\mathbf{r}_1\tau_1)\partial_\nu^1 G_0(\mathbf{r}_1\tau_1,\mathbf{r}_2\tau_2) \quad (B.3)$$

The Keldysh formalism

The solution to this equation is

$$\delta G(\mathbf{r}_1\tau_1, \mathbf{r}_2\tau_2) =$$
$$\sum_\nu \int d^3r_3 \int_{\tau_i}^{\tau_f} d\tau_3 t'(\tau_3) G_0(\mathbf{r}_1\tau_1, \mathbf{r}_3\tau_3)$$
$$\times \; (-\frac{i}{2}\partial_\nu^3(A_{s,\nu}(\mathbf{r}_3\tau_3)G_0(\mathbf{r}_3\tau_3, \mathbf{r}_2\tau_2)) - \frac{i}{2}A_{s,\nu}(\mathbf{r}_3\tau_3)\partial_\nu^3 G_0(\mathbf{r}_3\tau_3, \mathbf{r}_2\tau_2))$$
$$= \sum_\nu \int d^3r_3 \int_{\tau_i}^{\tau_f} d\tau_3 t'(\tau_3) G_0(\mathbf{r}_1\tau_1, \mathbf{r}_3\tau_3) \mathbf{j}_{op,\nu}(\mathbf{r}_3) G_0(\mathbf{r}_3\tau_3, \mathbf{r}_2\tau_2) \delta A_{s,\nu}(\mathbf{r}_3\tau_3)$$

as can readily be checked by insertion into the original equation (B.3) and using the equation of motion for G_0. We therefore obtain

$$\frac{\delta G(\mathbf{r}_1\tau_1, \mathbf{r}_2\tau_2)}{\delta A_\nu(\mathbf{r}_3, \tau_3)} = G_0(\mathbf{r}_1\tau_1, \mathbf{r}_3\tau_3)\mathbf{j}_{op,\nu}(\mathbf{r}_3)G_0(\mathbf{r}_3\tau_3, \mathbf{r}_2\tau_2) \quad (B.4)$$

Appendix: C

In this section we will describe derive some properties of product of Keldysh response functions and how they are transformed to real time response functions. All response functions are assumed to have the general structure

$$A(\mathbf{r}_1\tau_1, \mathbf{r}_2\tau_2) = A^\delta(\mathbf{r}_1, \mathbf{r}_2, t_1)\delta_C(t_1 - t_2) + \theta(\tau_1 - \tau_2)A^>(1, 2) + \theta(\tau_2 - \tau_1)A^<(1, 2) \quad (C.1)$$

where we use the short notation $1 = \mathbf{r}_1 t_1$ and where δ_C is the contour delta function $\delta_C(t_1 - t_2) = \delta(\tau_1 - \tau_2)/t'(\tau_1)$. For instance, for the current response function we have

$$\chi_{\mu\nu}^\delta = n_0(\mathbf{r}_1)\delta(\mathbf{r}_1 - \mathbf{r}_2)\delta_{\mu\nu} \quad (C.2)$$
$$\chi_{\mu\nu}^> = -i\langle\Psi_0|\Delta\mathbf{j}_{p,H,\mu}(1)\Delta\mathbf{j}_{p,H,\nu}(2)|\Psi_0\rangle \quad (C.3)$$
$$\chi_{\mu\nu}^< = -i\langle\Psi_0|\Delta\mathbf{j}_{p,H,\nu}(2)\Delta\mathbf{j}_{p,H,\mu}(1)|\Psi_0\rangle \quad (C.4)$$

We have already seen that if we let a Keldysh response function of the general form in Eq.(C.1) act on a perturbation $\delta v(\mathbf{r}_2 t(\tau_2))$ that is equal on both sides of the Keldysh contour that the reponse $\delta a(\mathbf{r}_1 t_1)$ isgiven as

$$\delta a(\mathbf{r}_1 t_1) = \int_C d^3 r_2 d\tau_2 t'(\tau_2) A(\mathbf{r}_1\tau_1, \mathbf{r}_2\tau_2)\delta v(\mathbf{r}_2 t_2)$$
$$= \int d^3 r_2 dt_2 A_R(\mathbf{r}_1 t_1, \mathbf{r}_2 t_2)\delta v(\mathbf{r}_2 t_2) \quad (C.5)$$

where the retarded response function A_R is now a function in real time given by

$$A_R(1, 2) = A^\delta(\mathbf{r}_1, \mathbf{r}_2, t_1)\delta(t_1 - t_2) + \theta(t_1 - t_2)(A^>(1, 2) - A^<(1, 2)) \quad (C.6)$$

Let now A and B be Keldysh response functions of the form given in Eq.(C.1). We then want to show that the function

$$(AB)(\mathbf{r}_1\tau_1, \mathbf{r}_2\tau_2) = \int d^3 r_3 d\tau_3 t'(\tau_3) A(\mathbf{r}_1\tau_1, \mathbf{r}_3\tau_3)B(\mathbf{r}_3\tau_3, \mathbf{r}_2\tau_2) \quad (C.7)$$

has the same general form Eq.(C.1) and that

$$(AB)_R(1, 2) = \int d3 A_R(1, 3) B_R(3, 2) \quad (C.8)$$

To show this we first write $A = a^\delta + a^r$ where a^δ and a^r are the singular and regular parts of A given by

$$a^\delta(\mathbf{r}_1\tau_1, \mathbf{r}_2\tau_2) = A^\delta(\mathbf{r}_1, \mathbf{r}_2, t_1)\delta_C(t_1 - t_2) \quad (C.9)$$
$$a^r(\mathbf{r}_1\tau_1, \mathbf{r}_2\tau_2) = \theta(\tau_1 - \tau_2)A^>(1,2) + \theta(\tau_2 - \tau_1)A^<(1,2) \quad (C.10)$$

With this notation the integral (C.7) has the form

$$(AB) = \int (a^\delta b^\delta + a^\delta b^r + b^\delta a^r + a^r b^r) \quad (C.11)$$

Let us start by evaluating the term $a^r b^r$. We have for the integral of $a^r b^r$ the following expression

$$\int_C d^3 r_3 d\tau_3 t'(\tau_3) a^r(\mathbf{r}_1\tau_1, \mathbf{r}_3\tau_3) b^r(\mathbf{r}_3\tau_3, \mathbf{r}_2\tau_2)$$
$$= \int_{\tau_i}^{\tau_f} d^3 r_3 d\tau_3 t'(\tau_3) \left(\theta(\tau_1 - \tau_3)\theta(\tau_3 - \tau_2)A^> B^> + \theta(\tau_1 - \tau_3)\theta(\tau_2 - \tau_3)A^> B^< \right.$$
$$+ \left. \theta(\tau_4 - \tau_3)\theta(\tau_4 - \tau_2)B^< C^> + \theta(\tau_4 - \tau_3)\theta(\tau_2 - \tau_4)B^< C^< \right)$$
$$= \theta(\tau_1 - \tau_2)(AB)^{r,>} + \theta(\tau_2 - \tau_1)(AB)^{r,<} \quad (C.12)$$

where

$$(AB)^{r,>} = \int d^3 r_3 \{ \int_{\tau_2}^{\tau_1} d\tau_3 t'(\tau_3) A^> B^> + \int_{\tau_i}^{\tau_2} d\tau_3 t'(\tau_3) A^> B^<$$
$$+ \int_{\tau_1}^{\tau_f} d\tau_3 t'(\tau_3) A^< B^> \} \quad (C.13)$$

$$(AB)^{r,<} = \int d^3 r_3 \{ \int_{\tau_1}^{\tau_2} d\tau_3 t'(\tau_3) A^< B^< + \int_{\tau_i}^{\tau_1} d\tau_3 t'(\tau_3) A^> B^<$$
$$+ \int_{\tau_2}^{\tau_f} d\tau_3 t'(\tau_3) A^< B^> \} \quad (C.14)$$

Since all terms under the integral depend on $t(\tau)$ they can be transformed to real time integrals. This yields

$$(AB)^{r,>} = \int d^3 r_3 \{ \int_{t_2}^{t_1} dt_3 A^> B^> + \int_{t_0}^{t_2} dt_3 A^> B^< - \int_{t_0}^{t_1} dt_3 A^< B^> \} \quad (C.15)$$
$$(AB)^{r,<} = \int d^3 r_3 \{ \int_{t_1}^{t_2} dt_3 A^< B^< + \int_{t_0}^{t_1} dt_3 A^> B^< - \int_{t_0}^{t_2} dt_3 A^< B^> \} \quad (C.16)$$

This is our first intermediate result. We continue to discuss the singular terms in Eq.(C.11). The term containing two equal time delta functions is easily evaluated to give

$$\int a^\delta b^\delta = \frac{\delta(\tau_1 - \tau_2)}{t'(\tau_1)} \int d^3 r_3 A^\delta(\mathbf{r}_1, \mathbf{r}_3, t_1) B^\delta(\mathbf{r}_3, \mathbf{r}_2, t_1) \quad (C.17)$$

This leaves us with the terms which contain one equal time delta function which are readily evaluated to be

$$\int (a^\delta b^r + a^r b^\delta) = \theta(\tau_1 - \tau_2)(AB)^{\delta,>} + \theta(\tau_2 - \tau_1)(AB)^{\delta,<} \quad (C.18)$$

where

$$(AB)^{\delta,>} = \int d^3 r_3 (A^\delta(\mathbf{r}_1, \mathbf{r}_3, t_1) B^>(\mathbf{r}_3 t_1, \mathbf{r}_2 t_2) + A^>(\mathbf{r}_1 t_1, \mathbf{r}_3 t_2) B^\delta(\mathbf{r}_3, \mathbf{r}_2, t_2)) \quad (C.19)$$

with an analogous equation for $(AB)^{\delta,<}$. If we combine our results we see that (AB) has the form

$$(AB)(\mathbf{r}_1 \tau_1, \mathbf{r}_2 \tau_2) = (AB)^\delta(\mathbf{r}_1, \mathbf{r}_2, t_1)\delta_C(t_1 - t_2) + \theta(\tau_1 - \tau_2)(AB)^>(1,2) \\
+ \theta(\tau_2 - \tau_1)(AB)^<(1,2) \quad (C.20)$$

where

$$(AB)^\delta(\mathbf{r}_1, \mathbf{r}_2, t_1) = \int d^3 r_3 A^\delta(\mathbf{r}_1, \mathbf{r}_3, t_1) B^\delta(\mathbf{r}_3, \mathbf{r}_2, t_1) \quad (C.21)$$

$$(AB)^> = (AB)^{r,>} + (AB)^{\delta,>} \quad (C.22)$$

$$(AB)^< = (AB)^{r,<} + (AB)^{\delta,<} \quad (C.23)$$

We see that the product (AB) has the same structure as the original functions A and B. The retarded product

$$(AB)_R(1,2) = (AB)^\delta(\mathbf{r}_1, \mathbf{r}_2, t_1)\delta(t_1 - t_2) + \theta(t_1 - t_2)((AB)^>(1,2) - (AB)^<(1,2)) \quad (C.24)$$

can now be evaluated. If we use

$$(AB)^{r,>} - (AB)^{r,<} = \int d^3 r_3 \int_{t_2}^{t_1} dt_3 (A^>(1,3) - A^<(1,3))(B^>(3,2) - B^<(3,2)) \quad (C.25)$$

we obtain

$$\begin{aligned}(AB)_R &= \int d^3 r_3 \int_{t_0}^{\infty} dt_3 (A^\delta \delta(t_1 - t_3) + \theta(t_1 - t_3)[A^>(1,3) - A^<(1,3)]) \\
&\quad \times (B^\delta \delta(t_3 - t_2) + \theta(t_3 - t_2)[B^>(3,2) - B^<(3,2)]) \\
&= \int d3 A_R(1,3) B_R(3,2) \quad (C.26)\end{aligned}$$

which proves our statement. By induction one can continue to prove the more general result

$$(A_1 A_2 \ldots A_n)_R(1,2) = \int d3 \ldots d(n+1) A_{1,R}(1,3) A_{2,R}(3,4) \ldots A_{n,R}(n+1,2) \quad (C.27)$$

Appendix: D

Here we demonstrate how the second order response function on the contour reduces to the causal one when acting on physical perturbations. The induced current has, up to second order, the expansion

$$
\begin{aligned}
\delta j_\mu(\bar{1}) &= \sum_\nu \int_C d2 \frac{\delta j_\mu(\bar{1})}{\delta A_\nu(\bar{2})} \delta A_\nu(\bar{2}) + \frac{1}{2} \sum_{\nu\kappa} \int_C d2 d3 \frac{\delta^2 j_\mu(\bar{1})}{\delta A_\nu(\bar{2})\delta A_\kappa(\bar{3})} \delta A_\nu(\bar{2}) \delta A_\kappa(\bar{3}) \\
&= \sum_\nu \int_C d2 \chi^{(1)}_{\mu\nu}(\bar{1},\bar{2}) \delta A_\nu(\bar{2}) \\
&+ \frac{1}{2} \sum_{\nu\kappa} \int_C d2 d3 \chi^{(2)}_{\mu\nu\kappa}(\bar{1},\bar{2},\bar{3}) \delta A_\nu(\bar{2}) \delta A_\kappa(\bar{3})
\end{aligned}
\quad \text{(D.1)}
$$

The second order response function has the form

$$
\begin{aligned}
\chi^{(2)}_{s,\mu\nu\kappa}(\bar{1},\bar{2},\bar{3}) &= (-i)^2 \langle T_C \Delta j_{p,H,\mu}(\bar{1}) \Delta j_{p,H,\nu}(\bar{2}) \Delta j_{p,H,\kappa}(\bar{3}) \rangle \\
&= \theta(\tau_1-\tau_2)\theta(\tau_2-\tau_3)\langle 123 \rangle + \theta(\tau_2-\tau_3)\theta(\tau_3-\tau_1)\langle 231 \rangle \\
&+ \theta(\tau_3-\tau_1)\theta(\tau_1-\tau_2)\langle 312 \rangle + \theta(\tau_1-\tau_3)\theta(\tau_3-\tau_2)\langle 132 \rangle \\
&+ \theta(\tau_3-\tau_2)\theta(\tau_2-\tau_1)\langle 321 \rangle + \theta(\tau_2-\tau_1)\theta(\tau_1-\tau_3)\langle 213 \rangle
\end{aligned}
\quad \text{(D.2)}
$$

where we introduced the short notation

$$
\langle ijk \rangle = (-i)^2 \langle \Delta j_{p,H,\mu}(\bar{i}) \Delta j_{p,H,\nu}(\bar{j}) \Delta j_{p,H,\kappa}(\bar{k}) \rangle
\quad \text{(D.3)}
$$

We then have for the second order change in the current (we use the short notation $2 = (\nu, \mathbf{r}_2, t_2(\tau_2))$ and $3 = (\kappa, \mathbf{r}_3, t_3(\tau_3))$ where $d2$ and $d3$ imply integration along the contour and summation over ν and κ) due to a perturbing field A:

$$
\begin{aligned}
2\delta j^{(2)}_\mu(1) &= \\
&\int_{\tau_i}^{\tau_1} d2 \int_{\tau_i}^{\tau_2} d3 \langle 123 \rangle A_2 A_3 + \int_{\tau_1}^{\tau_f} d3 \int_{\tau_3}^{\tau_f} d2 \langle 231 \rangle A_2 A_3 \\
&+ \int_{\tau_1}^{\tau_f} d3 \int_{\tau_i}^{\tau_1} d2 \langle 312 \rangle A_2 A_3 + \int_{\tau_i}^{\tau_1} d3 \int_{\tau_i}^{\tau_3} d2 \langle 132 \rangle A_2 A_3 \\
&+ \int_{\tau_1}^{\tau_f} d2 \int_{\tau_2}^{\tau_f} d3 \langle 321 \rangle A_2 A_3 + \int_{\tau_1}^{\tau_f} d2 \int_{\tau_i}^{\tau_1} d3 \langle 213 \rangle A_2 A_3 \\
&= \int_{t_0}^{t_1} d2 \int_{t_0}^{t_2} d3 \langle 123 \rangle A_2 A_3 + \int_{t_0}^{t_1} d2 \int_{t_0}^{t_2} d3 \langle 321 \rangle A_2 A_3 \\
&- \int_{t_0}^{t_1} d2 \int_{t_0}^{t_1} d3 \langle 213 \rangle A_2 A_3 + \int_{t_0}^{t_1} d2 \int_{t_0}^{t_2} d3 \langle 123 \rangle A_2 A_3 \\
&+ \int_{t_0}^{t_1} d2 \int_{t_0}^{t_2} d3 \langle 321 \rangle A_2 A_3 - \int_{t_0}^{t_1} d2 \int_{t_0}^{t_1} d3 \langle 213 \rangle A_2 A_3
\end{aligned}
\quad \text{(D.4)}
$$

where in some terms we interchanged the s 2 and 3 and use. If we use the expression

$$
\int_{t_0}^{t_1} d2 \int_{t_0}^{t_1} d3 \langle 213 \rangle A_2 A_3 = \int_{t_0}^{t_1} d2 \int_{t_0}^{t_2} d3 \langle 213 \rangle A_2 A_3 + \int_{t_0}^{t_1} d2 \int_{t_0}^{t_2} d3 \langle 312 \rangle A_2 A_3
\quad \text{(D.5)}
$$

(which follows by inspection of the integration regions and changing integration order and s) we obtain

$$
\begin{aligned}
2\delta j^{(2)}_\mu(1) &= 2 \int_{t_0}^{t_1} d2 \int_{t_0}^{t_3} d3 \langle 123 + 321 - 213 - 312 \rangle A_2 A_3 \\
&= 2 \int_{t_0}^{t_1} d2 \int_{t_0}^{t_3} d3 \langle [[1,2],3] \rangle A_2 A_3
\end{aligned}
\quad \text{(D.6)}
$$

This means that in physical time the second order response function is given by

$$\chi^{(2)}_{s,\mu\nu\kappa}(1,2,3) = (-i)^2 \theta(t_1-t_2)\theta(t_2-t_3)\langle\Psi_0|[[j_{p,H,\mu}(1),j_{p,H,\nu}(2)],j_{p,H,\kappa}(3)]|\Psi_0\rangle \tag{D.7}$$

which is the usual retarded second order response function. Note that taken out of the integrand this response function is not uniquely defined. We can, for instance, still symmetrize the function with respect to coordinates 2 and 3.

References

[1] V.Peuckert,J. Phys. C**11**, 4945, (1978)
[2] E.Runge and E.K.U.Gross, Phys. Rev. Lett. **52**, 997, (1984)
[3] E.K.U.Gross and W.Kohn, Adv.Quant.Chem. **21**, 255, (1990)
[4] E.K.U.Gross, J.F.Dobson and M.Petersilka,in *Topics in Current Chemistry*.(ed.) R.F.Nalewajski, Springer, (1996)
[5] R.van Leeuwen, Int.J.Mod.Phys.B**15**, 1969 (2001)
[6] R.van Leeuwen, Phys.Rev.Lett. **82**, 3863 (1999)
[7] N.T.Maitra and K.Burke, Phys.Rev. A**63**, 042501 (2001)
[8] X.Gonze, P.Ghosez and R.W.Godby, Phys.Rev.Lett. **74**, 4035 (1995)
[9] G.Vignale and W.Kohn, Phys.Rev.Lett. **77**, 2037 (1996)
[10] G.Vignale and W.Kohn, in *Electronic Density Functional Theory: Recent Progress and New Directions* eds. Dobson et al. (Plenum Press, New York, 1998)
[11] S.K.Ghosh and A.K.Dhara, Phys.Rev.A**38**, 1149 (1988)
[12] F.Kootstra, P.L.de Boeij and J.G.Snijders, J.Chem.Phys. **112**, 6517 (2000)
[13] F.Kootstra, P.L.de Boeij and J.G.Snijders, Phys.Rev. B**62**, 7071 (2000)
[14] G.F.Bertsch,J.-I.Iwata, A.Rubio and K.Yabana, Phys.Rev.B**62**, 7998 (2000)
[15] P.L.de Boeij, F.Kootstra, J.A.Berger, R.van Leeuwen and J.G.Snijders, J.Chem.Phys. **115**, 1995 (2001)
[16] C.A.Ullrich and G.Vignale, Phys.Rev.Lett. **87**, 037402 (2001)
[17] C.A.Ullrich and G.Vignale, Phys.Rev.B**65**, 245102 (2002)
[18] M.van Faassen, P.L.de Boeij, R.van Leeuwen, J.A.Berger and J.G.Snijders, Phys.Rev.Lett. **88**, 186401 (2002)
[19] R.van Leeuwen, Phys.Rev.Lett. **76**, 3610 (1996)
[20] R.van Leeuwen, Phys.Rev.Lett. **80**, 1280 (1998)
[21] J.Schwinger, J.Math.Phys. **2**, 407 (1961)
[22] L.V.Keldysh, Sov.Phys.JETP **20**, 1018 (1965)
[23] D.C.Langreth. *Linear and Nonlinear Electron Transport in Solids*,(eds.) J.T.Devreese and V.E.van Doren, Plenum, New York, 1976.
[24] R.Sandström.Phys. Stat. Sol. **38** 683, (1970)
[25] P.Danielewicz.Ann.Phys. **152**, 239, (1984)
[26] J.Rammer and H.Smith.Rev. Modern Phys., **58**, 323 (1986)
[27] A.L.Fetter and J.D.Walecka, *Quantum Theory of Many-Particle Systems* (McGraw-Hill, 1971)
[28] E.K.U.Gross, E.Runge and O.Heinonen, *Many-Particle Theory* (Adam Hilger, Bristol, 1991)
[29] M.Petersilka, U.J.Gossmann, and E.K.U.GrossPhys.Rev.Lett. **76**, 1212 (1996)
[30] C.A.Ullrich, U.J.Gossmann, and E.K.U.Gross.Phys. Rev. Lett., **74**, 872 (1995)
[31] A.Görling, Phys.Rev.A**55**, 2630 (1997)

[32] C.A.Ullrich, unpublished notes (1996)
[33] Z.Qian and G.Vignale, Phys.Rev.B**65**, 235121 (2002)
[34] P.Nozières and D.Pines, *The Theory of Quantum Liquids* (Perseus Books, 1999)
[35] J.Lindhard Det. Kongl. Danske Videnskab. Selsk. Mat.Fys.Med.**28**, no.8 (1954)
[36] A.Holas, P.K.Aravind and K.S.Singwi, Phys.Rev.B**20**, 4912 (1979)
[37] F.Brosens and J.T.Devreese, Phys.Rev.B**21**, 1363 (1980)
[38] S.Kurth and U.von Barth, to be published
[39] R.van Leeuwen, S.Kurth and U.von Barth, to be published

TOWARDS TIME-DEPENDENT DENSITY-FUNCTIONAL THEORY FOR MOLECULES IN STRONG LASER PULSES

T. Kreibich[1], N.I. Gidopoulos[2], R. van Leeuwen[3], and E.K.U. Gross[1]
[1] *Institut für Theoretische Physik, Freie Universität Berlin, Arnimallee 14, 14195 Berlin, Germany*
[2] *ISIS Facility, Rutherford Appleton Laboratory, Chilton, Didcot, Oxon, OX11 0QX, England, UK*
[3] *Theoretical Chemistry, Material Science Centre, Rijksuniversiteit Groningen, Nijenborgh 4, 9747 AG, Groningen, The Netherlands*

Abstract To describe the dynamical interplay of electronic and nuclear degrees of freedom in molecules exposed to strong laser pulses, we present two different variational approaches based on the statonary-action principle: A mean-field treatment of the electron-nuclear interaction and an explicitly correlated ansatz. The two methods are tested on a one-dimensional model of H_2^+ which can be solved exactly. The correlated approach significantly improves upon the mean-field treatment, especially in the case of laser fields strong enough to cause substantial dissociation.

1. Introduction

The past decade has witnessed rapid progress in laser technology. Nowadays, tabletop systems routinely provide femtosecond laser pulses with intensities in the terawatt regime. The field strengths at such intensities are comparable to or even larger than, typical atomic or molecular binding forces [1]. Therefore, an adequate description of strong-field multiphoton processes requires a non-perturbative scheme which treats the external laser field and the internal Coulomb forces of the atom or molecule on equal footing. While considerable progress has been made in understanding the behavior of atoms in high-intensity laser pulses, the situation for molecules is far less advanced since the additional nuclear degrees of freedom tremendously increase the complexity of the problem. The traditional methods like expanding the total molecular wavefunction in terms of few Born-Oppenheimer (BO) states or restricting oneself to a classical description of the nuclear degrees of freedom [2, 3] cannot satisfactorily explain the complex interplay between the electronic and the nuclear motion. In fact, the direct numerical solution of the full time-dependent (TD) Schrödinger equation (SE) for the H_2^+ molecular ion [4] shows that a proper treatment of all fundamental processes, i.e., electronic and vibrational excitation, ionization, and dissociation, is mandatory in the high-intensity regime. However, even for this smallest possible molecule, the full-scale numerical solution of the electron-nuclear TDSE is barely feasible. Clearly, the strong-field dynamics of larger molecules requires an approximate treatment. To this end, we employ the stationary-action principle

$$\delta \mathcal{A}[\Psi] = \delta \int_{t_0}^{t_1} dt \, \langle \Psi(t) | i \partial_t - \hat{H}(t) | \Psi(t) \rangle = 0 \qquad (1)$$

which, for any approximate form of the wavefunction Ψ, determines the corresponding equations of motion and thus the dynamical behavior of the system. In this contribution, we will present two different approximations for the total wavefunction Ψ: The first one is based on a mean-field-type treatment of the electron-nuclear coupling which neglects correlation effects between the electronic and the nuclear motion. The second approach employs an explicitly correlated ansatz for the electron-nuclear part of the wavefunction. The crucial point of both approaches is that they provide a description in terms of single-particle orbitals. Only single-particle equations, by their very nature, can be solved with moderate numerical effort and thus bring within reach the TD treatment of larger systems.

For simplicity, we restrict ourselves to the discussion of diatomic molecules consisting of N_e electrons and two nuclei of masses M_1 and M_2 and charges Z_1 and Z_2. Since we shall be interested in radiation sources in a regime where the dipole approximation holds true, the center-of-mass (CM) motion of the total system can be separated off. The molecule is then described by the vector of the internuclear distance $\mathbf{R}$ and electronic coordinates $\underline{\mathbf{r}} \equiv \{\mathbf{r}_j\}$ referring to a molecular body-fixed frame whose z-axis $\mathbf{e}_z$ is parallel to $\mathbf{R}$. In terms of these coordinates, the Hamiltonian in Eq. (1) reads (atomic units)

$$H(\underline{\mathbf{r}}, \mathbf{R}, t) = T(\underline{\mathbf{r}}, \mathbf{R}) + W(\underline{\mathbf{r}}, \mathbf{R}) + V_{\mathrm{L}}(\underline{\mathbf{r}}, \mathbf{R}, t) \qquad (2)$$

where $T = -(1/2\mu_n)\nabla_{\mathbf{R}}^2 - (1/2\mu_e)\sum_{j=1}^{N_e} \nabla_j^2$ with $\mu_n = (M_1 M_2)/(M_1 + M_2)$ and $\mu_e = (M_1 + M_2)/(M_1 + M_2 + 1)$ denotes the kinetic-energy operator, and $W = W_{ee} + W_{en} + W_{nn}$ contains the interactions between all particles. Mass-polarization and Coriolis terms are neglected. Furthermore, employing the length gauge, the laser potential in Eq. (2)

is given by $V_L(t) = \left(-q_n\mathbf{R} + q_e \sum_j \mathcal{R}(\mathbf{R})\mathbf{r}_j\right)\mathbf{E}(t)$, where $\mathbf{E}(t)$ denotes the electric field amplitude, $q_n = (Z_1 M_2 - Z_2 M_1)/(M_1 + M_2)$ and $q_e = (Z_1 + Z_2 + M_1 + M_2)/(M_1 + M_2 + N_e)$. $\mathcal{R}(\mathbf{R})$ represents the 3×3 rotational matrix which rotates the internuclear axis $\mathbf{R}$ (i.e., the z-axis of the body-fixed frame) into the z-axis of the CM-fixed coordinate system.

2. Formulation of DFT approach, absence of e-n correlation functional

In our first approach, the total electron-nuclear many-body wavefunction is approximated by

$$\Psi(\mathbf{R}, \underline{\mathbf{r}}, t) \approx \chi(\mathbf{R}, t)\psi(\underline{\mathbf{r}}, t), \tag{3}$$

where $\chi(\mathbf{R}, t)$ is a TD nuclear wavefunction and $\psi(\underline{\mathbf{r}}, t)$ represents a TD electronic many-particle state. The stationary-action principle (1) with Hamiltonian (2) then leads to a TDSE for the nuclear wavefunction

$$i\partial_t \chi(\mathbf{R}, t) = \left(-\frac{\nabla_\mathbf{R}^2}{2\mu_n} + V_{S,n}(\mathbf{R}, t)\right) \chi(\mathbf{R}, t) \tag{4}$$

and to a many-body TDSE for $\psi(\underline{\mathbf{r}}, t)$. Applying TD density-functional theory (TDDFT) [5, 6, 7] to the latter leads to a set of TD Kohn-Sham equations:

$$i\partial_t \varphi_j(\mathbf{r}, t) = \left(-\frac{\nabla_\mathbf{r}^2}{2\mu_e} + v_{S,e}(\mathbf{r}, t)\right) \varphi_j(\mathbf{r}, t), \tag{5}$$

where $\{\varphi_j(\mathbf{r}, t)\}$ are electronic Kohn-Sham orbitals. The local effective potentials $V_{S,n}$ and $v_{S,e}$ are given by

$$V_{S,n}(\mathbf{R}, t) = V_{L,n}(\mathbf{R}, t) + \frac{Z_1 Z_2}{R} + V_{H,n}^{en}(\mathbf{R}, t) \tag{6}$$

$$v_{S,e}(\mathbf{r}, t) = v_{L,e}(\mathbf{r}, t) + v_{Hxc}^{e}(\mathbf{r}, t) + v_{H,e}^{en}(\mathbf{r}, t). \tag{7}$$

The last terms on the r.h.s. of Eqs. (6) and (7) denote the time-dependent Hartree (mean-field) potentials arising from the electron-nuclear interaction. By virtue of the ansatz (3), they are given by the classical electrostatic potentials caused by the respective charge distributions:

$$V_{H,n}^{en}(\mathbf{R}, t) := \int d\mathbf{r}\, W_{en}(\mathbf{r}, \mathbf{R}) \rho(\mathbf{r}, t) \tag{8}$$

$$v_{H,e}^{en}(\mathbf{r}, t) := \int d\mathbf{R}\, W_{en}(\mathbf{r}, \mathbf{R}) \Gamma(\mathbf{R}, t), \tag{9}$$

where $\Gamma(\mathbf{R},t) = |\chi(\mathbf{R},t)|^2$ and $\rho(\mathbf{r},t) = \sum_{j=1}^{N_e} |\varphi_j(\mathbf{r},t)|^2$. The electron-nuclear interaction is given by $W_{en}(\mathbf{r},\mathbf{R}) = Z_1 \left|\mathbf{r} - \frac{M_2}{M_1+M_2}R\mathbf{e}_z\right|^{-1} + Z_2 \left|\mathbf{r} + \frac{M_1}{M_1+M_2}R\mathbf{e}_z\right|^{-1}$. The second term on the r.h.s. of Eq. (7) is the Hartree-exchange-correlation (Hxc) potential arising from the TDDFT treatment of the electronic degrees of freedom. Finally, the first terms on the r.h.s. of Eqs. (6) and (7) represent the laser potentials given by $V_{\mathrm{L},n}(\mathbf{R},t) = -q_n \mathbf{R}\mathbf{E}(t)$ and $v_{\mathrm{L},e}(\mathbf{r},t) = q_e \int d\mathbf{R}\, \Gamma(\mathbf{R},t)\, (\mathcal{R}(\mathbf{R})\mathbf{r})\, \mathbf{E}(t)$.

3. Ansatz to describe e-n correlation

In order to assess the accuracy of the scheme presented above, we apply it to a simplified model of the H_2^+ molecule which can be solved exactly. In this model, the dimensionality of the problem is reduced by confining the dynamics of all particles to one spatial dimension, i.e., the particles are allowed to move only in the direction of the laser polarization axis. Extensive studies have demonstrated that this model qualitatively reproduces all typical non-linear phenomena observed in strong-field experiments [8]. We emphasize that our primary interest is *not* the analysis of this model system. Instead, we aim at an ab-initio description of larger molecules exposed to strong laser fields. However, in order to assess the quality of the approximations employed, it is essential to have a numerically exact reference solution, to which the approximate results can be compared. The 1D model H_2^+ molecule naturally lends itself for that purpose.

The exact reference solution of the TDSE for the model H_2^+ molecule as well as the corresponding solution of the approximate equations of motion (4), (5) are obtained numerically by employing the split-operator technique [9]. The initial wavefunctions are chosen as the respective molecular ground states. As an example, we investigate the dynamics of the model H_2^+ molecule in a $\lambda = 770$ nm, 77 fs laser field, where the laser is linearly ramped from zero to its maximum value within 12.5 optical cycles and subsequently kept constant for another 17.5 optical cycles. Fig. 1 shows the time evolution of the expectation value of the internuclear separation $\langle R \rangle(t)$ for a peak laser intensity of $I_0 = 7.5 \times 10^{13}$ W/cm^2 as obtained from the exact solution (solid line) and from the approximate scheme (dashed line). As seen in this figure, the laser field causes the molecule to stretch in the beginning of the propagation. However, the field is not strong enough to substantially dissociate the system. Instead, the molecule starts to vibrate, which is reflected in the oscillatory structure of the mean internuclear distance $\langle R \rangle(t)$. Compared to the exact results, we find that, for the laser parameters considered here, the approximate scheme reproduces the main features of the dynamics: $\langle R \rangle(t)$

initially increases and subsequently oscillates, although on a quantitative level, the amplitude of the oscillations of $\langle R \rangle(t)$ is underestimated while its frequency is overestimated.

In order to analyze the origin of these deviations, we investigate the potential $V_{H,n}^{en}(\mathbf{R}, t)$ of the nuclear single-particle equation (4). This quantity is inherently time-dependent, but still it is instructive to consider the static nuclear potential resulting from the stationary ground-state solution of Eqs. (4) and (5) with the laser field switched off. In Fig. 2, this static nuclear ground-state potential is compared with the lowest-energy BO surface, which provides a very good reference for the ground state of the H_2^+ model molecule. The mean-field nuclear potential is in good agreement with the BO curve only in the neighborhood of the equilibrium internuclear separation. For larger values of R, the approximate nuclear potential severely deviates from the reference curve. ¿From this perspective, the deviations of the strong-field behavior are easily understood: Compared to the exact dynamics, the approximate scheme requires much more energy to stretch the molecule. Consequently, if the same laser parameters are used in the exact and the approximate calculation, the latter will significantly underestimate the nuclear motion, leading to the deviations found in $\langle R \rangle(t)$. Likewise, for laser parameters leading to substantial dissociation, the mean-field approach is found to underestimate the dissociation probability significantly as shown in Fig. 3. Hence, we conclude that a mean-field-type approximation for the electron-nuclear interaction cannot, provide a satisfactory picture of the strong-field dynamics of molecules.

In order to improve upon the mean-field approach, electron-nuclear correlation needs to be incorporated in the approximate form of the total wavefunction Ψ. We propose the explicitly correlated expression

$$\Psi(\mathbf{R}, \mathbf{r}, t) = \chi(\mathbf{R}, t)\, \varphi_{\mathbf{R}}(\mathbf{r}, t), \qquad (10)$$
$$\varphi_{\mathbf{R}}(\mathbf{r}, t) := \phi_1(\mathbf{r} - \frac{R}{2}\mathbf{e}_z, t) + \phi_2(\mathbf{r} + \frac{R}{2}\mathbf{e}_z, t)$$

where, $\chi(\mathbf{R}, t)$ again denotes the nuclear wavefunction. However, in contrast to the mean-field approach (3), the electronic degrees of freedom are not described by molecular orbitals, but by explicitly TD atomic orbitals $\phi_{1,2}(\mathbf{r}, t)$ which are attached to one of the nuclei. In other words, the correlation between the electron and the nuclei is introduced by referring the electron, in the spirit of a Heitler-London ansatz, to one or the other nucleus. The variationally best wavefunction of the form (10) is obtained by requiring the action to be stationary with respect to variations of all orbitals. This leads to [10] the equations of motion:

$$i\partial_t \chi(\mathbf{R},t) = \left(\hat{h}_n(\mathbf{R},t) - \Lambda(t)\right)\chi(\mathbf{R},t), \tag{11}$$

$$i\partial_t \phi_{1,2}(\mathbf{r},t) = \hat{h}_{e1,2}\,\phi_{1,2}(\mathbf{r},t) + \mathcal{Q}_{1,2}(\mathbf{r},t) \tag{12}$$

with the effective nuclear Hamiltonian

$$\hat{h}_n(\mathbf{R},t) := -\frac{1}{2\mu_n}\nabla_\mathbf{R}^2 - \frac{1}{\mu_n}\langle\nabla_\mathbf{R}\rangle_e(\mathbf{R},t)\,\nabla_\mathbf{R}$$
$$+\langle \hat{H} - i\partial_t\rangle_e(\mathbf{R},t). \tag{13}$$

and the effective electronic Hamiltonian

$$\hat{h}_{e1,2} := -\frac{1}{2\tilde{\mu}_e}\nabla_\mathbf{r}^2 \pm \frac{1}{2\mu_n}\langle\nabla_\mathbf{R}\rangle_n\,\nabla_\mathbf{r} \tag{14}$$
$$+\langle \hat{H}(\mathbf{r}\pm R/2\mathbf{e}_z,\mathbf{R}) - \hat{T}_e - i\partial_t\rangle_n$$

where $\tilde{\mu}_e = 4\mu_n\mu_e/(4\mu_n + \mu_e)$, and $\mathcal{Q}_{1,2}(\mathbf{r},t)$ denotes the inhomogeneity term : $\mathcal{Q}_{1,2}(\mathbf{r},t) := \langle(\hat{H}(\mathbf{r}\pm R/2\mathbf{e}_z,\mathbf{R}) - i\partial_t)\phi_{2/1}(\mathbf{r}\pm R\mathbf{e}_z,t)\rangle_n$ For ease of notation, the following abbreviations have been introduced: $\langle\hat{O}\rangle_e \equiv \langle\hat{O}\rangle_e(\mathbf{R},t) := \langle\varphi_\mathbf{R}|\hat{O}|\varphi_\mathbf{R}\rangle_e/\langle\varphi_\mathbf{R}|\varphi_\mathbf{R}\rangle_e$ and $\langle\hat{O}\rangle_n \equiv \langle\hat{O}\rangle_n(\mathbf{r},t) := \langle\chi|\hat{O}|\chi\rangle_n/\langle\chi|\chi\rangle_n$, where the subscripts "e" or "n" indicate the integration over the electronic or nuclear coordinate, respectively. Eqs. (11) and (12) form the heart of the new time-dependent variational scheme which, like the mean-field approach treats the strong external fields $V_{L,n}(t)$ and $v_{L,e}(t)$ and the intramolecular forces on the same footing in a non-perturbative way. Furthermore, the method properly accounts for the quantum nature of both the electronic and the nuclear degrees of freedom. In this respect, the proposed variational approach goes beyond the common mixed classical-quantum mechanical methods where the nuclear dynamics is treated classically. On the other hand, in contrast to methods employing the wavepacket propagation on few BO potential-energy surfaces, the influence of the strong laser field on the time evolution of the electrons is consistently incorporated as well. Still, although an explicitly correlated ansatz for the total wavefunction is used, it is important to realize that the dynamics is completely described in terms of single-particle orbitals. Consequently, the computational effort to solve Eqs. (11, 12) stays manageable. Considering these equations of motion individually, we observe the following features: The electronic equation (12) differs significantly from the corresponding mean-field equation (5). Whereas the latter describes the time evolution of molecular single-particle orbitals, the former propagates TD single-particle atomic orbitals. Consequently, additional inhomogeneity terms appear in Eq. (12)

which act as source or sink terms and are responsible for the (laser-induced) transfer of electronic charge between the two nuclei. Considering the effective electronic potentials, the contribution arising from the electron-nuclear interaction is given by $\langle W_{en}(\mathbf{r} \pm R/2\mathbf{e}_z, \mathbf{R})\rangle_n(\mathbf{r},t) = -1/r - (1/\langle\chi|\chi\rangle_n) \int d\mathbf{R}\ (|\chi(\mathbf{R},t)|^2/|\mathbf{r} \pm R\mathbf{e}_z|)$. Accordingly, the electron feels the bare Coulomb force of its reference nucleus and a Hartree-type potential from the second nucleus. In particular, due to the dependence of $\langle W_{en}\rangle$ on the time-dependent orbital $\chi(\mathbf{R},t)$, the formalism presented above naturally includes the quasistatic picture of molecular strong-field dynamics [11], which, e.g., successfully describes enhanced ionization leading to dissociative Coulomb explosion [12]. Turning to the nuclear equation of motion (11), we find that it is formally similar to Eq. (4). In particular, it again employs inherently TD effective potentials such that the nuclear dynamics is not restricted to a fixed potential-energy surface, but non-adiabatic processes can be described even by employing only one TD nuclear potential. The TD effective nuclear potential is explicitly given by the expectation value of $\hat{H} - i\partial_t$ with respect to $\varphi_\mathbf{R}(\mathbf{r},t)$, as seen from the last term of Eq. (12). Due to the Heitler-London-type form of $\varphi_\mathbf{R}$ which is correct in the asymptotic $(R \to \infty)$ limit the description of the effective nuclear potential and thus of the nuclear dynamics should be improved as compared to the TD mean-field scheme.

We employ again the same model H_2^+ to calculate the effective nuclear potential $V_n(R) = \langle \hat{H}_e \rangle_e(R)$ obtained from a self-consistent ground-state solution of the variational equations (11) and (12) (with the laser field switched off). Evidently, the resulting nuclear potential, as shown in Fig. 2, improves significantly upon the mean-field curve and only shows small deviations from the reference potential for $R \gtrsim 5$ a.u. Since the deviations of the nuclear mean-field potential were identified as the main source of error in the TD Hartree approach, the correlated variational scheme promises a more adequate treatment of the strong-field dynamics. This is confirmed by the results obtained for the dynamics of the model. The time evolution of the mean internuclear distance $\langle R\rangle(t)$, for the same laser parameters, is shown in Figs. 1,3 and the improvement is obvious. In Fig. 1 the agreement is excellent and even in the case of the strong field in Fig. 3, leading to considerable photodissociation, the qualitative picture is correct.

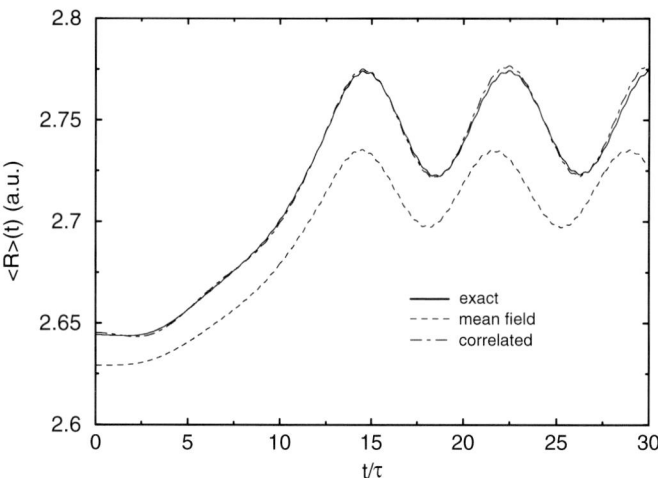

Figure 1. Time evolution (in units of the optical cycle τ) of the mean internuclear distance $\langle R \rangle(t)$ obtained for the model H_2^+ molecule, in a $\lambda = 770$ nm, $I_0 = 7.5 \times 10^{13}$ W/cm^2 laser field from the various methods described in the text.

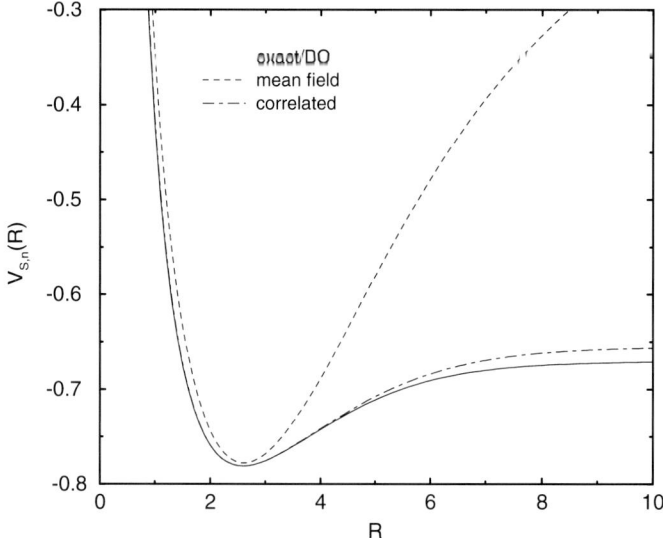

Figure 2. Effective nuclear potential $V_{S,n}(R)$ obtained for the model H_2^+ from the methods described. Atomic units.

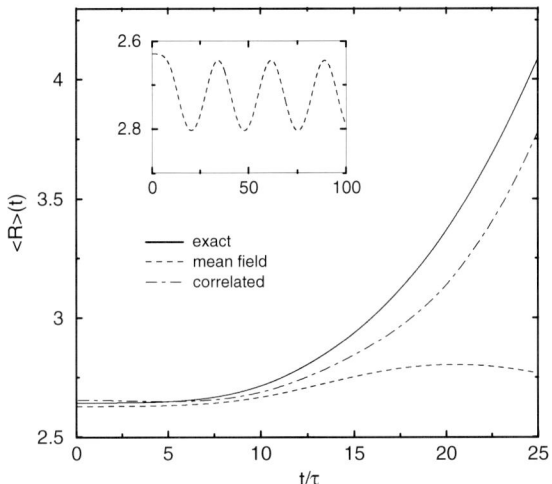

Figure 3. Time evolution of $\langle R \rangle(t)$ for the model H_2^+, in a $\lambda = 228$ nm, $I_0 = 2.5 \times 10^{13}$ W/cm^2 laser. In the insert, the mean field solution for a period of 100 τ.

References

[1] M. Protopapas, C.H. Keitel, and P.L. Knight, Rep. Prog.Phys. **60**, 389 (1997).
[2] U. Saalmann and R. Schmidt, Z. Phys. D **38**, 153 (1996); Phys. Rev. Lett. **80**, 3213 (1998).
[3] T. Kunert, R. Schmidt, Phys. Rev. Lett. **86**, 5258 (2001).
[4] S. Chelkowski, T. Zuo, O. Atabek, and A.D. Bandrauk, Phys. Rev. A **52**, 2977 (1995).
[5] E. Runge and E.K.U. Gross, Phys. Rev. Lett. **52**, 997 (1984).
[6] R. van Leeuwen, Phys. Rev. Lett. **80**, 1280 (1998); Phys. Rev. Lett. **82**, 3863 (1999).
[7] P. Hessler, J. Park and K. Burke, Phys. Rev. Lett. **82**, 378 (1999).
[8] Q. Su and J.H. Eberly, Phys. Rev. A **44**, 5997 (1991); R. Grobe and J.H. Eberly, Phys. Rev. Lett. **68**, 2905 (1992); W.-C. Liu and J.H. Eberly, S.L. Haan, R. Grobe, Phys. Rev. Lett. **83**, 520 (1999).
[9] M.D. Feit, J.A. Fleck Jr., and A. Steiger, J. Comput. Phys. **47**, 412 (1982).
[10] The ansatz (10) is invariant under $\chi \to c(t)\chi, \phi_i \to \phi_i/c(t)$. This freedom is fixed by introducing the complex purely TD function $\Lambda(t) = \langle \hat{h}_n \rangle_n(t)$.
[11] K. Codling and L.J. Frasinski, J. Phys. B: At. Mol. Opt. Phys. **26**, 783 (1993).
[12] S. Chelkowski and A.D. Bandrauk, J. Phys. B: At. Mol. Opt. Phys. **28**, L723 (1995).

PAIR DENSITY FUNCTIONAL THEORY

Á. Nagy
Department of Theoretical Physics, University of Debrecen,
H-4010 Debrecen, Hungary
anagy@madget.atomki.hu

Abstract   In the ground state the pair density can be determined by solving of an equation of a two-particle problem. The problem of an arbitrary system with even electrons is reduced to a two-particle problem. The effective potential of this two-particle equation has a term that is the functional derivative of the difference in the kinetic energies of the real system and an auxiliary system with respect to the pair density.

Keywords: density matrix, pair density, two-particle equation

1.   Introduction

The energy of an N-electron system may be determined exactly through a knowledge of the two-particle reduced density matrix [1, 2], because the interaction between the electrons are pairwise. However, the solution of the N-electron problem can only be reduced to the 2-particle density matrix problem if it is ensured that the 2-particle density matrix is derivable from an N-particle wave function. Unfortunatelly, no simple set of necessary and sufficient condition for the N-representability problem [3] is known.

Recently, the contracted Schrödinger equation is formulated [4, 5, 6], that resulted a new fundamental approach in density matrix theory.

In this paper another approach is presented. Following Gonis et al. [7] the 'hyperspace density' that is the diagonal of the spin independent reduced second-order density matrix or pair density is considered the key quantity. Recently, a density matrix functional theory is presented [8]. Now, another derivation is proposed. The approach of Theophilou [9] is generalized for pair density theory. This new formulation has the advantage that it bypasses the problem of v-representability. A different approach based on the pair density was proposed by Ziesche [10].

It is shown that in the ground state the pair density can be determined by solving of an equation of a two-particle problem. The problem of an arbitrary system with even electrons is reduced to a two-particle problem. The effective potential of this two-particle equation has an unknown term: the functional derivative of the difference in the kinetic energies of the real system and an auxiliary system with respect to the pair density.

2. Pair density functional theory

The Hamiltonian is generally written in the form

$$\hat{H} = \hat{T} + \hat{V}_{ee} + \sum_{i=1}^{N} v(\mathbf{r}_i), \qquad (1)$$

where $\hat{T}$ is the kinetic energy operator, $\hat{V}_{ee}$ is the electron-electron repulsion energy operator, and $v(\mathbf{r})$ is a local external potential. The kinetic energy and the electron-electron energy operators are

$$\hat{T} = \sum_{i=1}^{N} (-\frac{1}{2}\nabla_i^2) \qquad (2)$$

and

$$\hat{V}_{ee} = \sum_{i<j}^{N} \frac{1}{|\mathbf{r}_i - \mathbf{r}_j|} . \qquad (3)$$

Following Gonis et al. [7] we consider the Hamiltonian (1) but as a sum of distinct, nonoverlapping pairs of particles. It is assumed that the number of electrons is even: $N = 2M$. (A particle belongs to only one pair). As Gonis et al. proposed it is convenient to introduce six dimensional coordinate forms combining the two particles as a single one:

$$\mathbf{q} = (\mathbf{r}_i, \mathbf{r}_j) = (x_i, y_i, z_i, x_j, y_j, z_j) = (q_1, q_2, q_3, q_4, q_5, q_6) . \qquad (4)$$

Then the 'internal' potential for the particles in a pair $J = jj'$ is given by

$$\tilde{v}(\mathbf{r}_j, \mathbf{r}_{j'}) = \frac{1}{|\mathbf{r}_j - \mathbf{r}_{j'}|} . \qquad (5)$$

The interaction potential between the pairs has the form

$$\begin{aligned} W_{IJ} &= W(\mathbf{q}_I, \mathbf{q}_J) = W(\mathbf{r}_i, \mathbf{r}_{i'}; \mathbf{r}_j, \mathbf{r}_{j'}) \\ &= \frac{1}{|\mathbf{r}_i - \mathbf{r}_j|} + \frac{1}{|\mathbf{r}_i - \mathbf{r}_{j'}|} + \frac{1}{|\mathbf{r}_{i'} - \mathbf{r}_j|} + \frac{1}{|\mathbf{r}_{i'} - \mathbf{r}_{j'}|} . \end{aligned} \qquad (6)$$

The 'external' potential for the pair of particles is

$$\hat{U} = \sum_I u_I = \sum_{i \neq j}^{N} u(\mathbf{r}_i, \mathbf{r}_j) = v(\mathbf{r}_i) + v(\mathbf{r}_j) \,. \tag{7}$$

The Hamiltonian (1) can be partitioned in the following form:

$$\hat{H} = \hat{K} + \hat{W} + \hat{U} \,, \tag{8}$$

where

$$\hat{K} = \sum_{I=1} (-\frac{1}{2}\nabla_I^2 + \tilde{v}(\mathbf{q})) \,, \tag{9}$$

Denote by Φ an N-particle state in the Hilbert space with finite kinetic energy. That is Φ belongs to the space H^1 of square integrable functions whose derivatives are also square integrable. We assign the energy functional

$$K(\Phi) = Inf\left(\langle\Phi'|\hat{K}|\Phi'\rangle; n'(\mathbf{q}) = n(\mathbf{q}), \langle\Phi'|\Phi'\rangle = 1\right) \tag{10}$$

to each state. n is the diagonal of the spin-independent second-ordered density matrix n_2

$$n(\mathbf{r}_1, \mathbf{r}_2) = \int n_2(\mathbf{x}_1, \mathbf{x}_2; \mathbf{x}_1, \mathbf{x}_2) d\sigma_1 d\sigma_2$$
$$= \frac{N(N-1)}{2} \int |\Phi(\mathbf{x}_1, \mathbf{x}_2, \mathbf{x}_3..., \mathbf{x}_N)|^2 d\sigma_1 d\sigma_2 d\mathbf{x}_3...d\mathbf{x}_N \,. \tag{11}$$

$\mathbf{x}_i$ stands for the spatial and the spin coordinates $\mathbf{r}_i \sigma_i$
The following theorems hold.
THEOREM I

$$\langle\Phi|\hat{K}|\Phi\rangle \geq K(\Phi) \tag{12}$$

and for each Φ there is a Θ having the same n such that

$$\langle\Theta|\hat{K}|\Theta\rangle = K(\Phi) \tag{13}$$

Proof of Theorem I:
Lieb [4] proved that the infimum is a minimum for the three-dimentional density. (Theorem 3.3 in [4]) We expect that the proof would be the same for the six-dimentional pir density. That is for each Φ there exists a Θ with the property of Eq. (13).

THEOREM II

The variation of $\langle \Phi | \hat{K} | \Phi \rangle$ with respect to Φ under the pair density constraint leads the equation

$$\hat{K}\Phi + \hat{U}_{\text{eff}}\Phi = E'\Phi \tag{14}$$

Proof of Theorem II:

As the wave function should be normalized and n is kept fixed, we have to minimize the quantity $\langle \Phi | \hat{K} | \Phi \rangle - E' \langle \Phi | \Phi \rangle + \frac{1}{N-1} \int n(\mathbf{q}) u_{\text{eff}}(\mathbf{q})$, where E' and $u_{\text{eff}}(\mathbf{q})/(N-1)$ are the Lagrange multipliers. Taking into account the definition of n the variation of the last term is $\int \delta \Phi^* \sum_{i<j}^N u_{\text{eff}}(\mathbf{r}_i, \mathbf{r}_j)$. Then the variation gives:

$$\langle \delta \Phi^* | \hat{K} - E' + \hat{U}_{\text{eff}} | \Phi \rangle = 0, \tag{15}$$

where $U_{\text{eff}} = \sum_{i<j}^N u_{\text{eff}}(\mathbf{r}_i, \mathbf{r}_j)$ is a two-particle potential.

Then we define the functional L that contains the interaction energy between the electron pairs, too:

$$L(\Phi) = Inf \left(\langle \Phi' | \hat{K} + \hat{W} | \Phi' \rangle; n'(\mathbf{q}) = n(\mathbf{q}), \langle \Phi' | \Phi' \rangle = 1 \right) \tag{16}$$

The total energy functional is defined as

$$G(\Phi) = L(\Phi) + \langle \Phi | \hat{K} | \Phi \rangle - K(\Phi) + \frac{1}{N-1} \int n(\mathbf{q}) u(\mathbf{q}) d\mathbf{q} \tag{17}$$

THEOREM III

$$\langle \Phi | \hat{K} + \hat{W} | \Phi \rangle \geq L(\Phi) \tag{18}$$

and for each Φ there is a Ψ having the same n such that

$$\langle \Psi | \hat{K} + \hat{W} | \Psi \rangle = L(\Phi) \tag{19}$$

Proof of Theorem III:

From the analogue of Theorem 3.3 of Lieb [4], it follows that the infimum is a minimum. That is for each Φ there exists a Ψ that satisfies Eq. (19).

THEOREM IV

$$G(\Phi) \geq E_0, \tag{20}$$

where E_0 is the ground-state energy of the real system. If the minimum of $G(\Phi)$ is attained for some Φ_0, then the pair density corresponding to this Φ_0 is the correct ground-state pair density.

Proof of Theorem IV:
From the definition (17) of G and Theorem I follows that

$$G(\Phi) = L(\Phi) + \langle\Phi|\hat{K}|\Phi\rangle - K(\Phi) + \frac{1}{N-1}\int n(\mathbf{q})u(\mathbf{q})d\mathbf{q} \quad (21)$$

and

$$\langle\Phi|\hat{K}|\Phi\rangle \geq \langle\Theta|\hat{K}|\Theta\rangle . \quad (22)$$

Since

$$L(\Phi) + \frac{1}{N-1}\int n(\mathbf{q})u(\mathbf{q})d\mathbf{q} \geq E_0 , \quad (23)$$

we obtain Eq. (20) of theorem IV.

Consider now the ground-state wave function Ψ_0 and energy E_0 of the Hamiltonian H. The correponding ground-state pair function is n_0. Let Φ_0 be the wave function for which the minimum of $\langle\Phi'|\hat{K}|\Phi'\rangle$ is attained and for which the pair density is equal to n_0. From Eqs. (16) and (10) follows that

$$\begin{aligned} L(\Phi_0) &= Min\left(\langle\Phi'|\hat{K} + \hat{W}|\Phi'\rangle; n'(\mathbf{q}) = n_0(\mathbf{q}), \langle\Phi'|\Phi'\rangle = 1\right) \\ &= \langle\Psi_0|\hat{K} + \hat{W}|\Psi_0\rangle \end{aligned} \quad (24)$$

and

$$K(\Phi_0) = Min\left(\langle\Phi'|\hat{K}|\Phi'\rangle; n'(\mathbf{q}) = n_0(\mathbf{q}), \langle\Phi'|\Phi'\rangle = 1\right) . \quad (25)$$

The latter comes from the definition of Φ_0, while the former is the consequence of the fact that Ψ_0 is the ground-state wave function. (Note that $\Phi_0 = \Theta$ if both have the same pair density n_0.) From Eqs. (17), (24) and (25) we arrive at

$$\begin{aligned} G(\Phi_0) &= L(\Phi_0) + \langle\Phi_0|\hat{K}|\Phi_0\rangle - K(\Phi_0) + \frac{1}{N-1}\int n(\mathbf{q})u(\mathbf{q})d\mathbf{q} \\ &= \langle\Psi_0|\hat{K} + \hat{W} + \hat{U}|\Psi_0\rangle = E_0 , \end{aligned} \quad (26)$$

that is, the minimum of $G(\Phi)$ is the ground-state energy E_0.

Turning to the last part of the proof, let us suppose that for some Φ_0 we have $G(\Phi_0) = E_0$. The equality in the expression (20) holds if and only if (22) and (23) are equalities. Thus

$$L(\Phi_0) + \frac{1}{N-1}\int n(\mathbf{q})u(\mathbf{q})d\mathbf{q} = \langle\Psi|\hat{K} + \hat{W} + \hat{U}|\Psi\rangle = E_0 , \quad (27)$$

where Ψ is a state for which the pair density is the same as for Φ_0 (see Eq. (16).) But Eq. (27) implies that $\Psi = \Psi_0$ is the ground-state wave function of H and the ground-state pair density is the same as the pair density arising from Φ_0.

Now, define the functional $Z(\Phi)$

$$Z(\Phi) = L(\Phi) - K(\Phi) - \frac{N-2}{N-1} \int \frac{n(\mathbf{q})}{r_{12}} d\mathbf{q}, \tag{28}$$

where the last term in the right hand side of Eq.(28) is the expectation value of the operator $\hat{W}$: $\langle \Phi | \hat{W} | \Phi \rangle$ Then the total energy functional (Eq.(17)) can be written as

$$\begin{aligned} G(\Phi) &= \langle \Phi | \hat{K} | \Phi \rangle + Z(\Phi) + \frac{N-2}{N-1} \int \frac{n(\mathbf{q})}{r_{12}} d\mathbf{q} \\ &+ \frac{1}{N-1} \int n(\mathbf{q}) u(\mathbf{q}) d\mathbf{q}, \end{aligned} \tag{29}$$

Taking the variation of Eq.(28) with respect to Φ we are led to (Eq.(14)) with the potential

$$U_{\text{eff}} = \sum_{i<j}^{N} u_{\text{eff}}(\mathbf{r}_i, \mathbf{r}_j), \tag{30}$$

$$u_{\text{eff}} = \frac{1}{N-1} u(\mathbf{r}_i, \mathbf{r}_j) + \frac{N-2}{N-1} \frac{1}{|\mathbf{r}_i - \mathbf{r}_j|} + m(\mathbf{r}_i, \mathbf{r}_j), \tag{31}$$

where the last term is coming from the functional derivative of Z.

To write the solution of (Eq.(14)) we have to notice that the Hamiltonian is invariant with respect to the exchange of the coordinates of electrons within one pair but it is not invariant if the electrons in different pairs are exchanged. So the wave function can be written as a symmetrized expression of antisymmetric two-particle functions:

$$\Phi_0(\mathbf{x}_1, ..., \mathbf{x}_N) = \hat{S}(\chi_1(\mathbf{x}_1, \mathbf{x}_2)...\chi_M(\mathbf{x}_{N-1}, \mathbf{x}_N)). \tag{32}$$

$\hat{S}$ is the symmetrizer operator and the antisymmetric functions χ_I satisfy the equations

$$h^0(\mathbf{q})\chi_I(\mathbf{x}_1, \mathbf{x}_2) = \left[-\frac{1}{2}\nabla_{\mathbf{q}}^2 + \tilde{v}(\mathbf{q}) + u_{\text{eff}}(\mathbf{q}) \right] \chi_I(\mathbf{x}_1, \mathbf{x}_2) = \varepsilon_I \chi_I(\mathbf{x}_1, \mathbf{x}_2) \tag{33}$$

From Eq.(28) we see that the

$$T_c[n] = Z(\Phi) \tag{34}$$

is a functional of the pair density n. We assume that the functional derivative of T_c exists.

THEOREM V The antisymmetric two-particle functions are the solutions of

$$\left[-\frac{1}{2}\nabla_{\mathbf{q}}^2 + v_{\text{eff}}(\mathbf{q})\right]\chi_I(\mathbf{x}_1,\mathbf{x}_2) = \varepsilon_I\chi_I(\mathbf{x}_1,\mathbf{x}_2), \tag{35}$$

where

$$v_{eff} = v(\mathbf{r}_1) + v(\mathbf{r}_2) + \frac{N-1}{r_{12}} + v_k(\mathbf{q}), \tag{36}$$

$$v_k(\mathbf{q}) = (N-1)\frac{\delta T_c}{\delta n} \tag{37}$$

and

$$T_c = T - T^0 \tag{38}$$

is the difference of the kinetic energy of the real and auxiliary systems.

Proof of Theorem V:

To obtain the effective potential v_{eff} we rewrite Eq.(28) as follows. The first term in Eq.(28) has the form

$$\langle\Phi|\hat{K}|\Phi\rangle = T_0 + \frac{1}{N-1}\int\frac{n(\mathbf{q})}{r_{12}}d\mathbf{q}, \tag{39}$$

where

$$T^0 = \sum_{I=1}^{M}\int\chi_I^*(\mathbf{x}_1,\mathbf{x}_2)[-\frac{1}{2}\nabla_{\mathbf{q}}^2]\chi_I(\mathbf{x}_1,\mathbf{x}_2) \tag{40}$$

and the second term in Eq.(39) includes the total interaction energy inside the pairs. The pair density n can be written as

$$n(\mathbf{q}) = (N-1)\sum_{I=1}^{M}\sum_{\sigma}|\chi_I|^2. \tag{41}$$

So the total energy is given by

$$G(\Phi) = T^0 + T_c + \int\frac{n(\mathbf{q})}{r_{12}}d\mathbf{q} + \frac{1}{N-1}\int n(\mathbf{q})u(\mathbf{q})d\mathbf{q} \tag{42}$$

On the other hand, the total energy can be obtained as $\langle\Phi|\hat{H}|\Phi\rangle$ using Eq. (1). We can readily obtain

$$G(\Phi) = T + \int \frac{n(\mathbf{q})}{r_{12}} d\mathbf{q} + \frac{1}{N-1} \int n(\mathbf{q}) u(\mathbf{q}) d\mathbf{q}. \tag{43}$$

Comparison of Eqs.(42) and (43) immediatelly gives Eq.(38). The variation of Eq.(42) with respect to the two-particle functions leads to Eqs.(36) and (37).

The minimum energy is obtained if all two-particle functions χ_I are the same: $\chi_I = \chi_0$. Then

$$n(\mathbf{q}) = N\frac{N-1}{2}\sum_\sigma |\chi_0(\mathbf{x}_1,\mathbf{x}_2)|^2 = N\frac{N-1}{2}|\tilde{\chi}_0(\mathbf{q})|^2. \tag{44}$$

It means that the calculation of n is reduced to the solution of a single two-particle equation (35). So the N-body problem is reduced to a two-body problem.

3. Discussion

It is shown that in the ground state the pair density can be determined by solving an equation of a two-particle problem. The fact that the problem of an arbitrary system with even electrons can be reduced to a two-particle problem means enourmous simplification. It is found that the only unknown term in the effective potential of this two-particle equation is the functional derivative of the difference in the kinetic energies of the real system and an auxiliary system with respect to the pair density. An adequate enough approximation of this term would lead to a very simple treatment of an electron system. Unfortunatelly, T_c is not expected to be a small quantity, increasing roughly linearly with the number of electrons.

The present approach similar in methodology to the density functional theory [12]. In the density functional theory a non-interacting, or Kohn-Sham system [1] is studied instead of the real interacting system and the electrons are treated as if they move independently in a common effective local potential. Here, a 'non-interacting', auxiliary system is introduced and the electron pairs are treated as if they move independently in a common effective two-particle potential. The main difference is that the total wave function is not antisymmetric, but a symmetric expression of antisymmetric two-particle functions. It should also be noted that the idea of constructing wave functions as products of geminals is well-known [14].

The problem of v-representability is completely bypassed in the present formulation. The N-representability is also valid both in the real and the auxiliary systems. However, for an approximate functional T_c N-representability may not be satified and for an approximate n N-representability is not generally valid [15]. Consequently, the approximate form of T_c should be carefully chosen.

Acknowledgement

This work was supported by the grant OTKA No. T 029469.

References

[1] K. Husimi, Proc. Phys. Math. Soc. Jpn. **22**, 264 (1940).

[2] P. O. Löwdin, Phys. Rev. **97**, 1474 (1955).

[3] A. J. Coleman, Rev. Mod. Phys. 35, 668 (1963); E. R. Davidson, Reduced Density Matrices in Quantum Chemistry (Academic Press, New York, 1976); A. J. Coleman and V. I. Yukalov, Reduced Density Matrices: Coulson's Challange (Sringer-Verlag, New York, 2000); J. Cioslowski, Many-electron Densities and Reduced Density Matrices (Kluwer/Plenum, New York, 2000).

[4] H. Nakatsuji and K. Yasuda, Phys. Rev. Lett. **76**, 1039 (1996); K. Yasuda and H. Nakatsuji, Phys. Rev. A **56**, 2648 (1997).

[5] F. Colmenero and C. Valdemoro, Phys. Rev. A **47**, 979 (1993); C. Valdemoro, L. M. Tel and E. Perez-Romero, Adv. Quantum. Chem. **28**, 33 (1997).

[6] D. Mazziotti, Phys. Rev. A **57**, 4219 (1998); Chem. Phys. Lett. **289** 419 (1998); Int. J. Quantum. Chem. **70**, 557 (1998); Phys. Rev. A **60**, 3618 (1999); **60**, 4396 (1999).

[7] A. Gonis, T. C. Schulthess, J. van Ek and P. E. A. Turchi, Phys. Rev. Lett. **77**, 2981 (1996); A. Gonis, T. C. Schulthess, P. E. A. Turchi and J. van Ek, Phys. Rev. B **56**, 9335 (1997).

[8] Á. Nagy, Phys. Rev. A **66**, 022505 (2002).

[9] N. Hadjisavvas and A. Theophilou, Phys. Rev. A **30**, 2183 (1984); A. Theophilou, Int. J. Quantum Chem. **69**, 461 (1998).

[10] P. Ziesche, Phys. Lett. A **195**, 213 (1994); Int. J. Quantum. Chem. **60**, 149 (1996); M. Levy and P. Ziesche, J. Chem. Phys. **115**, 9110 (2001).

[11] E. Lieb, Int. J. Quantum. Chem. **24**, 243 (1983).

[12] P. Hohenberg and W. Kohn, Phys. Rev. **136**, B864 (1964).

[13] W. Kohn and L.J. Sham, Phys. Rev.**140 A**,1133(1965).

[14] J. M. Parks and R. G. Parr, J. Chem. Phys. **28**, 335 (1958); R. McWeeny, Proc. Roy. Soc. London A **253**, 242 (1959); T. A. Allen and H. Shull, J. Chem. Phys. **35**, 1644 (1961).

[15] E. R. Davidson, Chem. Phys. Lett. **246**, 209 (1995).

THE KUMMER VARIETY FOR N-PARTICLES

A. J. Coleman
Department of Mathematics and Statistics, Queen's University
Kingston, Ontario, Canada
colemana@post.queensu.ca

Abstract The Kummer Variety encapsulates the N-representability conditions, degeneracies, and phase transitions at zero temperature. A new method for calculating the energy of the ground state of a system of N fermions or bosons is suggested..

Keywords: N-representability conditions, Kummer Variety

1. Introduction

I suggest a new approach for calculating the energy of the ground state of a system of N fermions or bosons and, in telegraphic style, summarize the contents of my paper [1] in *Physical Review A* on: *"The Kummer Variety, the Geometry of N-representability, and Phase Transitions."*

The main results of this paper are contained in the paragraphs labelled **Basic Fact-III** and **Basic Fact-IV**. All references are to *"Reduced Density Matrices - Coulson's Challenge"* by A.J. Coleman and V.I. Yukalov [2] which will be cited as 'CY'. The key mathematical arguments are given in CY, sections 2.4, 2.5 and 2.6.

We are concerned with a system of N identical fermions or bosons studied in approximation using a finite one-particle basis set.

The theory is determined by:-

N, the number of particles,

r, the order of our 1-particle basis, $\{\varphi_i\}$,

s_n, the order of a basis of the linear space of n-particle functions which is equal to
$$\binom{r}{n}$$

for fermions,

and the choice of the spin-orbitals.

BASIC FACT I (BF-I)
All electrons are equal and none is more equal than others.

THUS. there are no such superior objects as s-, p-, d- electrons, superconducting electrons or Cooper pairs of electrons. Indeed, these would be contrary to British Common Law and to the US Constitution!

My friend, Dr. V. I.. Yukalov says that any thoughtful physicist knows this and regards such terms as what Bourbaki would call a mere innocent *"abus de language"* and that I tend to exaggerate the importance of this point. But I regard these terms as serious instances of Whitehead's *"Fallacy of Misplaced Concreteness"*. It is therefore mortal sin for teachers to use such terms with students. It may well be that this one habit among physicists and chemists explains the widespread difficulty in understanding the quantum mechanical N-body problem!

BASIC FACT II. (BF-II) (CY, p.11)

$$E = \frac{N}{2} tr\left(KD^2\right) = \frac{1}{2(N-1)} tr\left(K\rho_2\right)$$

where, the Reduced Hamiltonian (RH) is

$$K = H(1) + H(2) + (N-1)H(12)$$

and, in Dirac's normalization, the Second Order Reduced Density Operator, (*2-RDO*), is

$$\rho_2 = N(N-1)D^2.$$

2. Key Definitions

DEFINITION-I (CY, p. 56 & p. 63)
If B^2 is an arbitrary Hermitian operator on 2-particle space,

$$B^N := B^2 \wedge I^{N-2} = \Gamma_2^N B^2,$$

where $\wedge$ denotes the Grassmann Product of operators and Γ_2^N is the Kummer Expansion Operator which maps an operator on 2-space to one on N-space, so B^N is an Hermitian operator on N-fermion space.

Similarly, his Contraction Operator, L_N^2, maps an operator on N-space onto one on 2-space. These two operators are *adjoint* in the sense of (CY, Equ. 2.33); that is

$$\langle \Gamma_2^N B^2 | C^N \rangle_N = \langle B^2 | L_N^2 C^N \rangle_2,$$

where B^2 is an arbitrary operator on 2-space and C^N on N-space.

DEFINITION-II

The **KUMMER VARIETY** (KV) is the algebraic variety satisfying the equation

$$|B^2 \wedge I^{N-2}| = |B^N| = 0,$$

which is an equation of degree s_N in the elements of a matrix representing B^2. The set of Hermitian operators on 2-space of dimension s_2 can be thought of as a real linear space of dimension $(s_2)^2$. Since B^N is Hermitian, the equation,

$$\Delta(B^2) := |B^N| = 0,$$

has real coefficients and therefore corresponds to a variety of codimension unity.

DEFINITION-III

A *facet* of KV is a maximal connected subset of KV consisting of points B^2 at which $\Delta(B^2)$ has one simple zero, so B^N has nullity one.

If our system is condensed matter, we shall interpret the points of a facet of KV as parameters of one phase of the system.

DEFINITION-IV (CY, p.54)

The KUMMER CONE (KC) is the set of all B^2 such that B^N is positive semi-definite.

KC is a cone since any positive multiple of an element of the set is in the set. At an interior point all eigenvalues are positive; at a point on the boundary of KC, one or more are zero. The points of a facet are those for which exactly one eigenvalue is zero.

DEFINITION-V

A SHEET of KV is the set of points at which a particular eigenvalue of B^N is zero. Points on a sheet at which more than one eigenvalue is zero are not on a facet.

It is trivial to prove that I^2, the identity operator on 2-space, is interior to the KC. Therefore, no matter the direction of K, the PROBE, $I^2 + sK$, is interior to the KC for sufficiently small s.

3. Results

BASIC FACT III (BF-III) (CY, Equ. 2.49)
The zeros, s_i, of

$$\Delta(s) := \Delta\left(I^2 + sK\right) = |(I^2 + sK) \wedge I^{N-2}| = 0$$

are reciprocals of eigenvalues of the hamiltonian, H.

BASIC FACT IV (BF-IV) (CY, p.65)
At a point

$$B^2 = I^2 + s_i K$$

on a facet, the simple zero, s_i, corresponds to a wave function, ψ_i and is, apart from a multiplicative normalization factor, the directional (*functional*) derivative

$$\frac{\partial \Delta\left(B^2\right)}{\partial B^2} = \rho_2(\psi_i),$$

is the 2-RDM of the pure eigenstate, ψ_i, of H. Since all the components of the function on the left-hand side of this equation are homogeneous, it is trivial to satisfy the normalization condition,

$$tr(\rho_2) = N(N-1).$$

It follows from **BF-III** and **BF-IV** that if s_1 is the numerically smallest negative value of s such that $\Delta(s) = 0$, then $(s_1)^{-1}$ is the energy of the Ground State which is non-degenerate if the Probe is on a facet or, otherwise, degenerate.

A zero of $\Delta(s)$ of multiplicity μ will occur at an eigenvalue of H which is μ-*fold* degenerate and therefore at a point at which the boundaries of μ distinct facets meet.

The accompanying Figure illustrates the occurrence of DEGENERACIES and of PHASE TRANSITIONS AT ZERO TEMPERATURE. It purports to portray a cross-section through and orthogonal to I^2 and containing K which with no loss of generality we may assume satisfies $tr(K) = 0$. The triangle depicts the KC, the interiors of the sides o the triangle, facets of KV, the continuous lines, sheets of KV and the dashed line is the Probe emanating from I^2.

If the probe met the KV at a corner, the corresponding energy would be degenerate. By adding a perturbation to the Hamiltonian - for example, by taking account of **L** · **S** coupling - the direction of K would

The Kummer Variety for N-particles 93

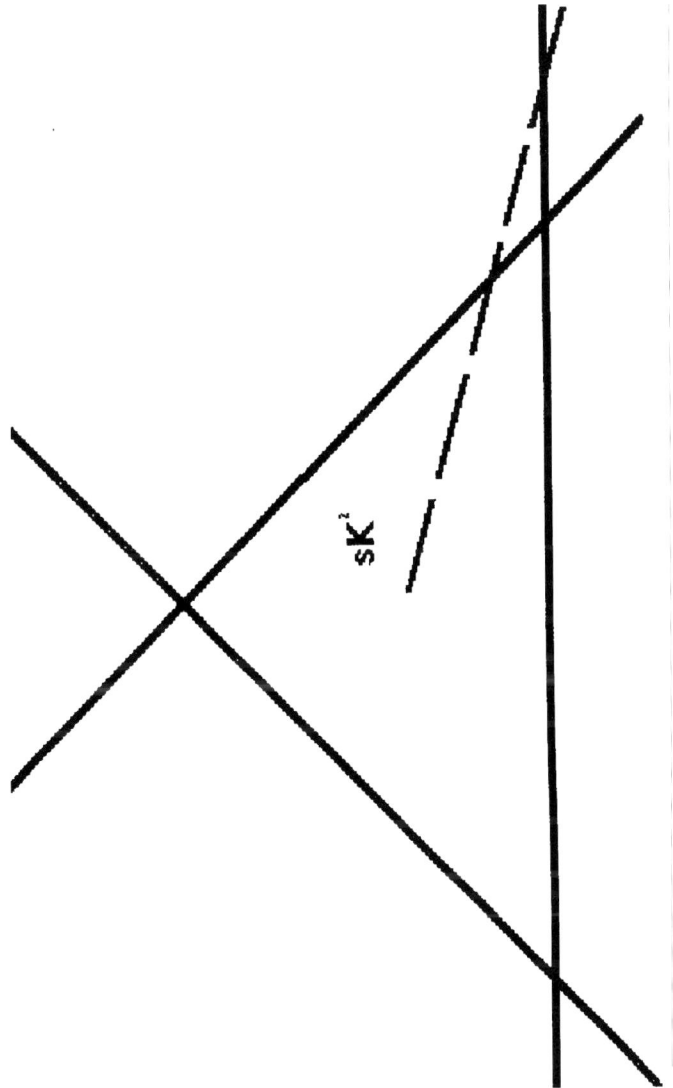

Figure 1. Two-dimensional cross section through I^2, orthogonal to the central axis, of part of the KV with the probe meeting the KC and a second sheet of KV.

change slightly and the degeneracy would be resolved into two, or more, levels.

Similarly, if the sides of the triangle depicted phases of a system of condensed matter, as the Probe passes through the singularity - perhaps in response to variation in a continuous parameter such as magnetic field - the direction of the normal to the facet might jump abruptly giving rise to a jump from one ρ_2 to another causing a phase transition.

To discuss coexistence of two phases, or systems at finite temperature such as Type II superconductors, would require RDM theory for mixed states. The general outline of how this could be done appears obvious but the details would take us beyond the scope of the present paper.

4. Next Steps?

Since $\Delta(s) = 0$, the key equation of our theory is of the same degree as the basic equation of Full Configuration Interaction (FCI) theory, at first glance, our theory does not seem to represent any essential advance compared with FCI.

One difference is that FCI seeks the s_N coefficients of the Wave Function, we seek the $(s_2)^2$ coefficients of ρ_2 and for these we give the formula of **BF-IV** for them. It should be possible to write an off-the-shelf program which, given s_i for an eigenvalue of ψ_i and the one-particle basis, outputs $\rho_2(\psi_i)$.

It is therefore my hope that progress can be made along one or more of the following lines:-

1. The value for the Probe at s_1 for the Ground State is written in terms of the physically meaningful Second Order Reduced Hamiltonian. It might therefore be easier than with FCI to truncate the matrix for B^N while not decreasing accuracy too greatly, but reducing the size of the sparse determinants to be evaluated.

2. The recent paper by Mazziotti and Erdahl [3] which marks a very important step forward in using lower-bound methods to approximate ground state energy, suggests that, since it involves rather considerable effort, it might be worthwhile to compare its expense with the following method.

    (a) Use the lower-bound procedure as it stood in 1975, surveyed for example in Garrod et al.. [4], to obtain a moderately good lower-bound, then

(b) sharpen that bound, by interpolation or other methods, using our formula for $\Delta(s)$.

This method would lead to equally sharp upper *and* lower bounds for s_1, while avoiding the initial problem of FCI of obtaining the eigenvalue of a huge sparse matrix. Its value compared with other available procedures can probably be evaluated only by numerical exploration which I would encourage.

3 It would be desirable to discover methods of working directly with geminals and, of course, of finding a set of natural geminals (nags) since the 2-RDM is completely determined by its s_2 nags!

4 An idea that I have played with for two or more decades, but so far with no practical result, is that we should study the asymptotic form of the solution of CY, Eqn. (7.11) as N increases without limit in the light of discussion on *pp.* 255-258.

5 I suspect that some of the difficulty of the N-particle problem arises from the role of the finite number N and that possibly much of it will disappear for N equal to infinity. If so, it might be possible to find sets of relatively simple geminals which would provide a reasonable approximation for the nags of a system with large N.

This may sound like wild speculation, which, indeed, it is. But are we not in a situation where our only option is that of Martin Luther: *"Sin boldly!"*?

References

[1] A.J. Coleman, *Phys. Rev. A* 66, 022503 (2002)
[2] A.J. Coleman and V.I. Yukalov, *"Reduced Density Matrices - Coulson's Challenge"*, Lecture Notes in Chemistry, Springer Verlag, Berlin (2000)
[3] Mazziotti and Erdahl, *Phys. Rev. A* **63**, 04213 (2001)
[4] Garrod et al.. *J.Math.Phys.* **16**, 923-930 (1975)

SOME UNSOLVED PROBLEMS IN DENSITY MATRIX THEORY AND DENSITY FUNCTIONAL THEORY

Roy McWeeny
Dipartimento di Chimica e Chimica Industriale
Via Risorgimento 35, 56100 Pisa, Italy.
mcweeny@dcci.unipi.it

Abstract Density functional theory (DFT) has its roots in density matrix theory (DMT). We therefore startwith a brief survey of the 1- and 2-electron density matrices (1-DM and 2-DM) and the densities derive from them, along with their main properties.

The problem of how to pass from the DMs to an energy density functional is considered in the light of simple examples: these show the richness of structure hidden within the 1-DM. In DFT, the diagonal element of the spinless 1-DM (i.e. the electron density, P) is deemed sufficient to determine everything else, including the 2-DM and the related electron interaction energy. In DMT, the 1-DM and 2-DM are indeed sufficient. But in either case the densities must be subject to the condition that they derive from an acceptable N-electron wave function – they must be *N-representable*. Otherwise, variational calculations will 'collapse', giving meaningless results.

In the absence of a general procedure for imposing N-representability, one may use an *ansatz* for the densities – forms which are *known* to arise (by integration) from an N-electron function of some suitable approximate form. An IPM approximation, with a 1-determinant wave function, is easy to handle but is too limited.

Procedures for obtaining the DMs are studied. The most general are essentially group theoretical and involve the separation of space and spin variables: one starts from the the Schrödinger equation $\mathsf{H}\Phi = E\Phi$, asking which solutions are physically admissible (most of them will not be) and what conditions they bring with them.

The availability of an N-representable 2-DM appears to be the minimum requirement for deriving an acceptable electron density and avoiding all reference to an N-electron wave function.

1. Preliminaries: the density functions

In 1964 Hohenberg and Kohn [1] showed that the total electronic energy of an N-electron system may, at least in principle, be determined as a functional of the electron density[1]; and the search for the form of such a functional still continues. Here we use concrete examples to touch upon some of the fundamental aspects of density functional theory (DFT).

First, it should be stressed that the HK theorem is an *existence* theorem: it does not lead to any prescription for actually calculating the energy. Many years ago, E. Breit Wilson proposed another 'proof' of the theorem

> Given an electron density function, you look for the cusps: these give you the positions of any nuclei (at density peaks) and the nuclear charges (from the cusp forms), while integration of the density tells you the number of electrons, N. Then you can set up the N-electron Schrödinger equation; and on solving it you get a set of eigenvalues (E_K) and eigenfunctions (Ψ_K). The lowest value of E_K is the exact ground-state energy, which is thus a *functional* of the electron density! From the corresponding eigenfunction, simply by quadrature, you can get all other properties of the state – so these are also functionals of the given density.

As he rightly remarked, this is not very helpful! The great boom in DFT during the last 30 years has arisen from recipes proposed by Kohn and Sham [2] and many others (for a bibliography see Ref.[3]) for the practical construction of *approximate* energy functionals – recipes that lean heavily on conventional Hartree-Fock theory.

Let us begin with a summary of definitions and concepts (a more complete development is available elsewhere [4]). As everyone knows, the most general definition of the electron density is[2]

$$P(\mathbf{r}) = N \int \Psi(\mathbf{x}, \mathbf{x}_2, ... \mathbf{x}_N)\Psi^*(\mathbf{x}, \mathbf{x}_2, ... \mathbf{x}_N) ds d\mathbf{x}_2 ... d\mathbf{x}_N \qquad (1)$$

– which, more correctly, is the probability/unit volume of finding an electron (no matter which) at point $\mathbf{r}$. In this definition, the wavefunction (in principle exact) is assumed antisymmetric and the spin variables are eliminated in the integrations. The factor N arises because, *whichever* electron is at point $\mathbf{x}$, integration over the other $N-1$ variables must give an identical result ($\Psi\Psi^*$ being symmetric).

The quantity $P(\mathbf{r})$ is in fact the 'diagonal element', $P(\mathbf{r}) = P(\mathbf{r}; \mathbf{r})$, of the (spinless) 1-electron density *matrix* $P(\mathbf{r}; \mathbf{r}')$, in which the $\mathbf{r}$ of Ψ^*

---
[1] Also called the 'number density' (probable number of electrons per unit volume) or the 'charge density' (probable amount of electric charge per unit volume, in units of $-e$).
[2] $\mathbf{x}$ will be used to denote space and spin variables collectively, $\mathbf{x} = \mathbf{r}, s$

in (1) is replaced by a second variable $\mathbf{r}'$. And $P(\mathbf{r}; \mathbf{r}')$ in turn is derived from a density matrix $\rho(\mathbf{x}; \mathbf{x}')$ with spin included [3]:

$$\rho(\mathbf{x}; \mathbf{x}') = N \int \Psi(\mathbf{x}, \mathbf{x}_2, \ldots \mathbf{x}_N) \Psi^*(\mathbf{x}', \mathbf{x}_2, \ldots \mathbf{x}_N) d\mathbf{x}_2 \ldots d\mathbf{x}_N \quad (2)$$

from which, identifying s', s and integrating,

$$P(\mathbf{r}; \mathbf{r}') = \int_{s'=s} \rho(\mathbf{x}; \mathbf{x}') ds. \quad (3)$$

A key property of this density matrix is that the expectation value of any symmetric sum of 1-electron operators, such as the electron-nuclear potential energy (V_{en}) and the electronic kinetic energy (T) in the Hamiltonian, can be written directly in terms of the *one*-electron quantity. Thus, with the 'standard' (non-relativistic, clamped-nuclei) Hamiltonian

$$\mathsf{H} = \mathsf{T} + V_{en} + V_{ee} = \sum_{i=1}^{N} \frac{\mathsf{p}^2(i)}{2m} + \sum_{i=1}^{N} V(\mathbf{r}_i) + \sum_{i \neq j=1}^{N} g(i,j), \quad (4)$$

the electron-nuclear potential energy and the kinetic energy terms yield expectation values

$$V_{en} = \int V(\mathbf{r}) P(\mathbf{r}) d\mathbf{r}, \ (a) \qquad \langle \mathsf{T} \rangle = \int_{\mathbf{r}'=\mathbf{r}} \frac{\mathsf{p}^2}{2m} P(\mathbf{r}; \mathbf{r}') d\mathbf{r}, \ (b) \quad (5)$$

where the off-diagonal quantity is required only for the *kinetic* energy. The potential energy is thus a very simple functional of the electron density, with an obviously 'classical' interpretation; but the kinetic energy is a functional of the density *matrix* and has so far resisted all attempts to express it in terms of the electron density alone - except in the trivial case of a free-electron gas. This difficulty is fundamental, arising from the non-commuting nature of position and momentum operators, and in most practical applications of DFT it is now customary to express $\langle \mathsf{T} \rangle$ as in (5) i.e. as a functional of the density *matrix* rather than the density. We note that the expectation value $\langle \mathsf{T} \rangle$ is obtained essentially by a *trace* operation – putting $\mathbf{r}' \to \mathbf{r}$ (i.e. taking a 'diagonal element') and then integrating over $\mathbf{r}$ (i.e. taking a 'diagonal sum'). Thus, with an alternative notation,

$$\langle \mathsf{T} \rangle = \mathrm{tr} \left(\frac{\mathsf{p}^2}{2m} \right) P(\mathbf{r}; \mathbf{r}'), \quad (6)$$

'tr' denoting the 'integral trace'.

Since $P(\mathbf{r};\mathbf{r}')$ arises by spin integration in (3), it is important to know the general form of $\rho(\mathbf{x};\mathbf{x}')$. In fact,

$$\begin{aligned}\rho(\mathbf{x};\mathbf{x}') &= P_{\alpha,\alpha}(\mathbf{r};\mathbf{r}')\alpha(s)\alpha^*(s') + P_{\alpha,\beta}(\mathbf{r};\mathbf{r}')\alpha(s)\beta^*(s') \\ &+ P_{\beta,\alpha}(\mathbf{r};\mathbf{r}')\beta(s)\alpha^*(s') + P_{\beta,\beta}(\mathbf{r};\mathbf{r}')\beta(s)\beta^*(s'),\end{aligned} \quad (7)$$

this being the integral-operator form of the density *operator* ρ, whose effect on a space-spin function is given by $\rho\psi(\mathbf{x}) = \int \rho(\mathbf{x};\mathbf{x}')\psi(\mathbf{x}')d\mathbf{x}'$. For any state of definite spin, however, only the α,α and β,β components are non-zero. On using (3),

$$P(\mathbf{r};\mathbf{r}') = P_{\alpha,\alpha}(\mathbf{r};\mathbf{r}') + P_{\beta,\beta}(\mathbf{r};\mathbf{r}') \quad (8)$$

and on using the analogue of (5), with spin included, to evaluate the expectation value of the total spin z component, we obtain

$$\langle \mathsf{S}_z \rangle = \int_{\mathbf{x}'=\mathbf{x}} \mathsf{S}_z \rho(\mathbf{x};\mathbf{x}')d\mathbf{x} = \int Q_z(\mathbf{r};\mathbf{r})d\mathbf{r}, \quad (9)$$

where, generally,

$$Q_z(\mathbf{r};\mathbf{r}') = \tfrac{1}{2}[P_{\alpha,\alpha}(\mathbf{r};\mathbf{r}') - P_{\beta,\beta}(\mathbf{r};\mathbf{r}')] \quad (10)$$

and has a diagonal element[3], usually referred to simply as "the spin density" [5], which is the density of spin angular momentum around the z axis. Thus

$$\begin{aligned} P(\mathbf{r}) &= P_\alpha(\mathbf{r}) + P_\beta(\mathbf{r}), & (11) \\ Q_z(\mathbf{r}) &= \tfrac{1}{2}[P_\alpha(\mathbf{r}) - P_\beta(\mathbf{r})] & (12) \end{aligned}$$

are, respectively, the electron density and the *spin density*. It must be emphasized at this point, however, that Q_z is only one component of a *vector* density[4], the field components $Q_x(\mathbf{r}), Q_y(\mathbf{r})$ being defined in strict analogy with (8). Attention is usually focused on Q_z (z being the quantization axis) only because for a pure spin state (quantum numbers S, M) the components transverse to the axis are everywhere zero. For such a state, $Q_x = Q_y = 0$ and $Q_z = (M/S)Q_z^{\text{st}}$, where Q_z^{st} denotes the density for a standard state – the 'top' state ($M = S$) of the multiplet. This is not the case in the presence of spin-orbit coupling, as will be seen later.

---

[3]In referring to diagonal elements we drop both the second variable and the corresponding spin index. The resultant density is often normalized to unity (dividing by M).
[4]More correctly a *pseudo*vector density, being an angular momentum per unit volume.

There is another spatial density function of great importance – the *current density* $\mathbf{J}(\mathbf{r})$, which is again a vector density. The components $J_x(\mathbf{r}), J_y(\mathbf{r}), J_z(\mathbf{r})$ determine the flux of electron (probability) density out of the volume element $d\mathbf{r}$ at any point $\mathbf{r}$ in space. Since the flux is zero everywhere for a state described by a *real* wavefunction, it is necessary in discussing this density to introduce in the Hamiltonian a magnetic field – however weak – in order to define the quantization axis and to introduce an imaginary component in Ψ. The presence of such a field requires only that the usual kinetic energy term in the Hamiltonian, namely $(1/2m)\sum_i \mathsf{p}^2(i)$, be replaced by

$$\mathsf{T} = (1/2m)\sum_i \pi^2(i), \qquad \pi_\lambda = \mathsf{p}_\lambda + eA_\lambda \qquad (\lambda = x, y, z). \tag{13}$$

Here π is the 'gauge invariant momentum' and contains the vector potential $\mathbf{A}$ of the field: π is in fact the operator associated with the electron *velocity*. The components of the current density are defined most symmetrically as

$$J_\lambda(\mathbf{r}) = m^{-1}[\tfrac{1}{2}(\pi_\lambda + \pi_\lambda'^\dagger)P(\mathbf{r};\mathbf{r}')]_{\mathbf{r}'=\mathbf{r}}, \tag{14}$$

where the adjoint operator $\pi_\lambda'^\dagger$ works on the *primed* variables in the density matrix.

The diagonal element $J_\lambda(\mathbf{r})$, has all the properties of a *local* flux density component at point $\mathbf{r}$, as defined in classical physics [6]. In particular, for a wavefunction satisfying the time-dependent Schrödinger equation, it is easy to derive a conservation equation for the electron density:

$$\mathrm{div}\,\mathbf{J}(\mathbf{r}) = -\frac{\partial P(\mathbf{r})}{\partial t} \tag{15}$$

– the rate of decrease of the probability of finding an electron in volume element $d\mathbf{r}$ at point $\mathbf{r}$ is the divergence of the current density at that point. It follows that for a system in a stationary state the currents must be *circulating* ($\mathrm{div}\,\mathbf{J} = 0$); and that these induced currents (created by the external field) will produce a secondary magnetic field. The components of this field, at point $\mathbf{r}_0$, are found to be

$$B_\lambda^{\mathrm{ind}}(\mathbf{r}_0) = \left(\frac{\mu_0}{4\pi}\right) \int \frac{[(\mathbf{r}-\mathbf{r}_0) \times -e\mathbf{J}(\mathbf{r})]_\lambda}{|\mathbf{r}-\mathbf{r}_0|^3} d\mathbf{r} \tag{16}$$

– just as if (remembering the Biot-Savart law) the electron distribution were a conducting medium containing electric currents of density $-e\mathbf{J}(\mathbf{r})$. Another interesting property of the current density is that the total electronic energy of an arbitrary system in the presence of any external field whatever, with vector potential **A**, may be written [6]

$$E = E_0 - \int \mathbf{A} \cdot [-e\mathbf{J}_0(\mathbf{r})]d\mathbf{r} - \tfrac{1}{2}\int \mathbf{A} \cdot [-e\mathbf{J}_{\text{ind}}(\mathbf{r})]d\mathbf{r}. \qquad (17)$$

Here E_0 and $\mathbf{J}_0$ are the energy and current density in the zero-field limit ($\mathbf{J}_0$ vanishing except for systems of high symmetry in states with unquenched angular momentum e.g. paramagnetic atoms), while $\mathbf{J}_{\text{ind}}$ is the field-linear part of the current density and therefore describes the currents *induced* by the applied field. The two field-dependent terms in (17) represent, respectively, the potential energy of a system of 'permanent' electric currents, of density $-e\mathbf{J}_0(\mathbf{r})$, and that of a system of induced currents, of density $-e\mathbf{J}_{\text{ind}}(\mathbf{r})$, in a conducting medium in the presence of a field – as would be calculated from classical physics.

To complete the formulation of an expression for the total electronic energy, we need to include electron repulsion terms: the expectation value of this energy, which is a symmetric sum $\tfrac{1}{2}\sum_{i,j} g(i,j)$, follows in terms of the *two*-electron DM (2-DM) $\pi(\mathbf{x}_1, \mathbf{x}_2; \mathbf{x}'_1, \mathbf{x}'_2)$ and takes the form

$$V_{ee} = \tfrac{1}{2}\text{tr}(1)\text{tr}(2)g(1,2)\pi(\mathbf{x}_1, \mathbf{x}_2; \mathbf{x}'_1, \mathbf{x}'_2)$$

– subscripts 1,2 now labelling *two* points in space. The 2DM $\pi(\mathbf{x}_1, \mathbf{x}_2; \mathbf{x}'_1, \mathbf{x}'_2)$ is the precise analogue of the 1-DM $\rho(\mathbf{x}; \mathbf{x}')$. When $g(1,2)$ involves neither differential nor spin operators (being the usual function of interelectron distance), primed variables may be dropped and we may also integrate immediately over spins. The result is

$$V_{ee} = \tfrac{1}{2}\int g(1,2)\Pi(\mathbf{r}_1, \mathbf{r}_2)d\mathbf{r}_1 d\mathbf{r}_2, \qquad (18)$$

where $\Pi(\mathbf{r}_1, \mathbf{r}_2)$ is the diagonal element, or 'pair density', analogous to the electron density $P(\mathbf{r}_1)$ in the 1-electron case.

The function $\pi(\mathbf{x}_1, \mathbf{x}_2; \mathbf{x}'_1, \mathbf{x}'_2)$ may be written in a form resembling (7), but with 16 components instead of 4: of these, for a state of definite spin, only 6 are non-zero [7], and (integrating over spins) only 4 give contributions to $\Pi(\mathbf{r}_1, \mathbf{r}_2; \mathbf{r}'_1 \mathbf{r}'_2)$. This *spinless* pair density has a diagonal element

$$\Pi(\mathbf{r}_1, \mathbf{r}_2) = \Pi_{\alpha\alpha}(\mathbf{r}_1, \mathbf{r}_2) + \Pi_{\alpha\beta}(\mathbf{r}_1, \mathbf{r}_2) + \Pi_{\beta\alpha}(\mathbf{r}_1, \mathbf{r}_2) + \Pi_{\beta\beta}(\mathbf{r}_1, \mathbf{r}_2), \qquad (19)$$

whose terms indicate, respectively, the probability that the electrons at points $\mathbf{r}_1$ and $\mathbf{r}_2$ are both 'up-spin', one 'up-spin' and one 'down-spin', or both 'down-spin'.

The importance of the density functions is that the total electronic energy may be written, without approximation (apart from the use of a clamped-nuclei non-relativistic Hamiltonian), in the form[5]

$$E = T + V_{en} + V_{ee}, \qquad (20)$$

where the kinetic energy, electron-nuclear potential energy, and electron-electron potential energy are given in (5(a)), (5(b)), and (18), respectively. In summary,

$$\begin{aligned}
T &= (1/2m)\int_{\mathbf{r}'=\mathbf{r}} \mathbf{p}\cdot\mathbf{p}'^{\dagger} P(\mathbf{r};\mathbf{r}')d\mathbf{r}, \\
V_{en} &= \int V(\mathbf{r})P(\mathbf{r})d\mathbf{r}, \\
V_{ee} &= \tfrac{1}{2}\int g(1,2)\Pi(\mathbf{r}_1,\mathbf{r}_2)d\mathbf{r}_1 d\mathbf{r}_2.
\end{aligned} \qquad (21)$$

This physically transparent form of the energy expression is indeed the starting point for DFT – where the aim is, ideally, to eliminate both $P(\mathbf{r};\mathbf{r}')$ and $\Pi(\mathbf{r}_1,\mathbf{r}_2)$ in favour of the electron density alone. An apparently more modest aim would be to use the equations (21) as they stand, with suitable approximate forms of the densities, as a basis for variational calculations of the energy – *without reference to the wavefunction itself*: how to do so constitutes the famous 'N-representability problem' which was recognized [8] even before DFT, in the 1950s, and remains unsolved – except in a somewhat cryptic sense (see Ref.[9] for a recent survey), not lending itself easily to practical application.

2. Density matrices to density functionals

The expressions in (5) and (18) expose the main problems of DFT. Only V_{en} is an explicit functional of the electron density $P(\mathbf{r})$. The kinetic energy is an explicit functional of the density *matrix*; and the major part of V_{ee} is also explicit, being the 'self-energy'

$$V_{se} = \tfrac{1}{2}\int g(1,2)P(\mathbf{r}_1)P(\mathbf{r}_2)d\mathbf{r}_1 d\mathbf{r}_2 \qquad (22)$$

of the charge distribution. But the rest is as yet unknown and is usually absorbed into an 'exchange-correlation' functional E_{xc}. Only the second term in (20) involves explicitly the external potential $V(\mathbf{r})$ ('external' in the sense that it specifies the field in which the electrons move and is thus characteristic of the particular system considered): the remaining part of E should therefore be a *universal* functional of the density, derivable from the density itself without reference to the nuclear framework of the system – and the same should evidently be true of E_{xc}. In current

---

[5] Note that T is expressed in a symmetrical form, the vector operators $\mathbf{p}, \mathbf{p}'$ working on the unprimed and primed variables, respectively, and the dagger indicating the adjoint. In the presence of a magnetic field $\mathbf{p}$ should be replaced by $\boldsymbol{\pi}$, its 'gauge invariant' counterpart.

DFT much effort is devoted to finding plausible semi-empirical forms of the exchange-correlation functional. In DMT, on the other hand, the energy is explicitly determined as a functional of the 2-DM and there is no need to introduce semi-empirical expressions: the only snag is that the precise form of Π is not known. It is true that any acceptable 'trial' functionals may be used in approximating the terms in (20), thus providing a basis for variational calculations; but they must somehow embody the conditions of N-representability – they must be obtainable from some legitimate approximate wave function.

In the absence of necessary and sufficient conditions for the N-representability of density matrices (see for example Refs.[8] - [10]), one may start from a rather general *ansatz*. The 1-DM arising from a pure-state wave function expressed in configuration interaction (CI) form can always be represented [11] as

$$\rho(\mathbf{x};\mathbf{x}') = \sum \lambda_r \psi_r(\mathbf{x}) \psi_r^*(\mathbf{x}'), \quad 0 \leq \lambda_r \leq 1, \quad \sum_r \lambda_r = N, \qquad (23)$$

where the 'spin-orbitals' (not referred to at all until now!) are simply the functions that diagonalize the 1-DM and are called the 'natural' spin-orbitals (NSOs). They are eigenfunctions of the 1-DM, which may in principle be *exact*, and therefore satisfy $\rho \psi_r(\mathbf{x}) = \lambda_r \psi_r(\mathbf{x})$, or more fully

$$\int \rho(\mathbf{x};\mathbf{x}') \psi_r(\mathbf{x}') = \lambda_r \psi_r(\mathbf{x}), \qquad (24)$$

where the eigenvalue λ_r is the 'occupation number' of ψ_r in the CI wave function. The expansion (23) is in principle infinite but in practice is always truncated.

The pair DM $\pi(\mathbf{x}_1, \mathbf{x}_2; \mathbf{x}_1', \mathbf{x}_2')$ may similarly be expressed in terms of natural spin 'geminals' (NSGs); but again the NSGs must be derived from a wave function [10][12] and their use has been limited.

In one particular case, an N-representable pair density can be constructed explicitly. This case corresponds to the truncated form of (23) in which

$$\rho(\mathbf{x};\mathbf{x}') = \sum_{r=1}^{N} \psi_r(\mathbf{x}) \psi_r^*(\mathbf{x}'), \qquad (25)$$

which is the 'Fock-Dirac' density matrix. This quantity exhibits the projection operator property of idempotency ($\rho\rho = \rho$) $\int \rho(\mathbf{x};\mathbf{x}'') \rho(\mathbf{x}'';\mathbf{x}') d\mathbf{x}'' = \rho(\mathbf{x};\mathbf{x}')$; and this property itself is sufficient to guarantee that the density derives from a 1-determinant wave function, the spin-orbitals being the N eigenfunctions of ρ with unit eigenvalues. As is well known [11], the corresponding density matrices for any number of electrons are all functionals of $\rho(\mathbf{x};\mathbf{x}')$: for example, the 2-DM is

$$\pi(\mathbf{x}_1,\mathbf{x}_2;\mathbf{x}_1',\mathbf{x}_2') = \begin{vmatrix} \rho(\mathbf{x}_1;\mathbf{x}_1') & \rho(\mathbf{x}_1;\mathbf{x}_2') \\ \rho(\mathbf{x}_2;\mathbf{x}_1') & \rho(\mathbf{x}_2;\mathbf{x}_2') \end{vmatrix} \quad (26)$$

and the n-DM is a determinant of the same form. Consequently all the terms in (20) become functionals of $\rho(\mathbf{x};\mathbf{x}')$ and variational procedures based on the Fock-Dirac density matrix [13], which make no reference to the wavefunction itself, are therefore soundly based and will give correct upper bounds to the energy. Unfortunately, this is true only at the Hartree-Fock level of approximation.

The more general density matrix (23) is *not* idempotent; and there is no known way of building from it the 2-electron density $\pi(\mathbf{x}_1,\mathbf{x}_2;\mathbf{x}_1',\mathbf{x}_2')$ which determines the electron repulsion energy. In this event, whatever approximate form may be assumed for π, which determines all the terms in (21), variational calculations based on (20) – though perfectly possible – cannot be guaranteed free of variational collapse. Nor can the results of such calculations have any guaranteed physical meaning within the framework of orthodox quantum mechanics. To illustrate the nature of this dilemma, it is sufficient to consider two elementary examples.

3. Examples: density functions for simple systems

To illustrate the practical calculation of the various density functions we consider two elementary examples, using wave functions of IPM form.

3.1 Density functions for the Boron atom

In IPM approximation the ground-state electron configuration is assumed to be

$$B[1s^2 2s^2 2p^1]$$

In the absence of spin-orbit coupling, any state is a simultaneous eigenstate of the mutually commuting operators $\mathsf{H}_0, \mathsf{L}^2, \mathsf{L}_z, \mathsf{S}^2, \mathsf{S}_z$: such states are characterized by E and the quantum numbers L, M_L, S, M_S. The ground state is here 6-fold degenerate, with energy $E = 2\epsilon_{1s} + 2\epsilon_{2s} + \epsilon_{2p}$; and the density matrix contains a closed-shell term common to all states, giving zero contributions to both the spin and current densities, plus a p-electron contribution (which varies according to the values of M_L, M_S). We therefore consider only the p-shell density contributions. The 'states' specified by quantum numbers M_L, M_S are not of course true spectroscopic states: the Hamiltonian is invariant under common space and spin rotations and we should be looking for simultaneous eigenfunctions of J^2, J_z. In the limit of zero spin-orbit coupling the 'correct zero-order' eigenfunctions can still be written in 1-determinant form; but the p-electron will occupy a 2-component spin-orbital $\psi = a\phi^\alpha \alpha + b\phi^\beta \beta$. The density contributions are easily found: omitting the common closed-shell term, they are as shown in Table 1.

Table 1. Densities for the Boron atom (L-S coupling)

J, M_J	Charge	Spin	Current
$\frac{3}{2}, \frac{3}{2}$	$P = \frac{1}{2}(x^2 + y^2)f^2$ (Torus)	$Q_x = 0$ $Q_y = 0$ $Q_z = \frac{1}{2}P$	$J_x = -yf^2$ $J_y = xf^2$ $J_z = 0$
$\frac{3}{2}, \frac{1}{2}$	$P = \frac{1}{6}(4z^2 + x^2 + y^2)f^2$ (Ellipsoid)	$Q_x = -\frac{2}{3}xzf^2$ $Q_y = -\frac{2}{3}yzf^2$ $Q_z = \frac{1}{12}(4z^2 - x^2 - y^2)f^2$	$J_x = -\frac{1}{3}yf^2$ $J_y = \frac{1}{3}xf^2$ $J_z = 0$
$\frac{1}{2}, \frac{1}{2}$	$P = \frac{1}{3}(x^2 + y^2 + z^2)f^2$ (Sphere)	$Q_x = \frac{2}{3}xzf^2$ $Q_y = \frac{2}{3}yzf^2$ $Q_z = \frac{1}{6}(z^2 - x^2 - y^2)f^2$	$J_x = -\frac{2}{3}yf^2$ $J_y = \frac{2}{3}xf^2$ $J_z = 0$

Only contributions from the p-shell are shown in Table 1: the $1s^2 2s^2$ 'core' contributes a spherically symmetric charge density, but zero spin and current densities. The 2p orbitals (real form) are $xf(r), yf(r), zf(r)$. When L-S coupling is assumed J, M_J are the appropriate quantum numbers.

It is to be noted that changing the sign of M_J reverses the direction of the spin polarizaton and the sense of the current flow. In the absence of electron interaction all states would have the same energy, namely $E = 2\epsilon_{1s} + 2\epsilon_{2s} + \epsilon_{2p}$.

The richness of structure in the various density functions is not evident on inspecting only the electron density $P(\mathbf{r})$; for it depends on the eigenvalues of the maximal set of mutually commuting operators required in specifying the state uniquely. This structure disappears if we take an ensemble average (thus renouncing all knowledge of observables other than the energy): the ensemble electron density then becomes spherically symmetrical and the spin and current densities disappear.

3.2 A 2-electron system: singlet and triplet

Again in IPM approximation, let us consider 2 electrons in a degenerate pair of orbitals ϕ_1, ϕ_2. The spins may be coupled to singlet or triplet and in the absence of electron interaction the states are degenerate. This time we calculate the 2-body quantity $\pi(\mathbf{x}_1, \mathbf{x}_2; \mathbf{x}'_1, \mathbf{x}'_2)$ for the states with $S = 1, M_S = 0$ and $S = M_S = 0$, obtaining (with spin variables in the usual order s_1, s_2, s'_1, s'_2)

Singlet: $\pi_S = F \times (\alpha\beta\alpha^*\beta^* - \alpha\beta\beta^*\alpha^* - \beta\alpha\alpha^*\beta^* + d\beta\alpha\beta^*\alpha^*)$.
Triplet: $\pi_T = G \times (\alpha\beta\alpha^*\beta^* + \alpha\beta\beta^*\alpha^* + \beta\alpha\alpha^*\beta^* + \beta\alpha\beta^*\alpha^*)$.

Here F and G are linear combinations of orbital products, symmetric and antisymmetric, respectively, under interchange of spatial variables $\mathbf{r}_1 \leftrightarrow \mathbf{r}_2$ and $\mathbf{r}'_1 \leftrightarrow \mathbf{r}'_2$. Integration over spins leads to

$$\Pi_S(\mathbf{r}_1, \mathbf{r}_2; \mathbf{r}'_1, \mathbf{r}'_2) = 2F(\mathbf{r}_1, \mathbf{r}_2; \mathbf{r}'_1, \mathbf{r}'_2) \qquad (27)$$
$$\Pi_T(\mathbf{r}_1, \mathbf{r}_2; \mathbf{r}'_1, \mathbf{r}'_2) = 2G(\mathbf{r}_1, \mathbf{r}_2; \mathbf{r}'_1, \mathbf{r}'_2); \qquad (28)$$

and a further spatial integration leads to

$$P_S(\mathbf{r}_1; \mathbf{r}'_1) = P_T(\mathbf{r}_1; \mathbf{r}'_1) = \phi_1(\mathbf{r}_1)\phi_1^*(\mathbf{r}'_1) + \phi_2(\mathbf{r}_1)\phi_2^*(\mathbf{r}'_1)$$

– giving a density which is the same for both singlet and triplet. It is because the *two*-electron densities are very different in the two cases that admission of electron interaction breaks the degeneracy. In fact, inspection of the various terms in Π (which describes the *correlation* of electronic motions) shows that (i) no two volume elements may simultaneously contain electrons of the same spin ($\Pi_{\alpha\alpha}$ and $\Pi_{\beta\beta}$ being everywhere zero); and (ii) while electrons of different spin may be found in the same volume element ($\mathbf{r}_2 \to \mathbf{r}_1$) in the *singlet* state, this is forbidden in the *triplet* state – the double antisymmetry of G ensuring that $\Pi_{\alpha\beta}(\mathbf{r}_1, \mathbf{r}_2) \to 0$ for $\mathbf{r}_2 \to \mathbf{r}_1$ like r_{12}^2 (exactly as in the 'Fermi hole' for electrons of the *same* spin). The Examples above confirm the dubious

nature of any arguments based only on 'up-spin' and 'down-spin' interactions: the physically significant interactions are related to the *coupling* of spins (e.g. to the spin scalar product $\mathsf{S}(i) \cdot \mathsf{S}(j)$) and the strength of this coupling, for electrons at any two points in space, is measured by a 'spin-coupling density' [8] – and this is derived from the *pair* density matrix. In fact, the terms 'up-spin' and 'down-spin' electrons are simplistic abstractions which do not offer any basis for a 'two-fluid' model, with N_α electrons of one kind and N_β of the other, even though such models are commonly invoked.

They also show that, even if knowledge of the electron density may in principle be sufficient to determine a state of lowest energy, it cannot (in the presence of degeneracy) determine *which* state – unless other information is provided. Different states are characterized by the different eigenvalues of *any other operators that may commute with the Hamiltonian*: thus, Example (i) shows that for an electron in a central field reversal of an angular momentum component will change the current or spin density *without changing the charge density* – whose form depends, however, on the total angular momentum (either L alone or, in the presence of L-S coupling, the resultant J); while Example (ii) shows that, for two electrons in *any* potential field, the different forms of the *pair* density are determined by the spin coupling (singlet or triplet) but that the difference is obliterated in the integration that leads to the electron density – so that, knowing only the latter, one cannot identify the state. In general, on passing from the pair density to the electron density, information is irrevocably lost; and the same is true at all levels as one passes from the N-electron density matrix ($\Psi\Psi^*$) to the reduced density matrices.

It might be objected that the examples considered are 'artificial' and that the ground state does not normally comprise a degenerate set of simultaneous eigenfunctions of several commuting operators. But this objection cannot be sustained since, as will become clear in the next Section, even the *exact* eigenfunctions of a spin-free Hamiltonian are in general highly degenerate.

Ideally, the aim of both DMT and DFT would be to eliminate all reference to the N-electron wavefunction, working entirely in terms of the 1- and 2-electron density functions P and Π. In the extreme case of DFT, P alone would be deemed sufficient; but from now on we consider the more general goal of getting the 2-DM, $\Pi(\mathbf{r}_1, \mathbf{r}_2; \mathbf{r}'_1 \mathbf{r}'_2)$, from which P follows by quadrature. The ground-state energy would certainly follow rigorously from Π; but both DMT and DFT have a common problem, which can be stated as follows:

> Given some 'trial' density $\tilde{\Pi}$, how is it possible to 'build in' the requirement that this function derive from some wave function $\tilde{\Phi}$ that is a legitimate approximation to the exact ground-state eigenfunction?

To shed light on this problem, it is useful to consider first the properties that an exact wave function must possess.

4. A group theoretical approach

It is important to remember that electron *spin*, unlike the classical position and momentum variables, is in a certain sense extraneous to DFT: the Hamiltonian operator most commonly employed in electronic structure calculations, namely (4), does not include spin operators or any terms of relativistic origin – which are usually introduced subsequently (if at all) as small corrections to the non-relativistic energy. In non-relativistic DFT it is in principle quite unnecessary to introduce electron spin explicitly: one might simply assume that the Schrödinger equation

Some Unsolved Problems

H$\Phi = E\Phi$ *could* be solved exactly and then ask which of the solutions are physically admissible (many of them will not be) and whether they would impose constraints on the reduced density matrices.

Let us suppose then that exact eigenfunctions of the Hamiltonian are available, namely functions $\Phi_\kappa(\mathbf{r}_1, \mathbf{r}_2, \ldots \mathbf{r}_N)$, containing only *spatial* variables and satisfying the partial differential equation

$$H\Phi_\kappa(\mathbf{r}_1, \mathbf{r}_2, \ldots \mathbf{r}_N) = E\Phi_\kappa(\mathbf{r}_1, \mathbf{r}_2, \ldots \mathbf{r}_N) \quad (29)$$

where, for generality, we may consider a set of g degenerate solutions with a common energy $E_1 = E_2 = \ldots = E_g = E$. Such solutions normally exhibit a complicated behaviour under electron permutations, but relatively few of these exact eigenfunctions are physically acceptable: the rest must be thrown away! It is the *symmetry* that determines *which* eigenfunctions can be used in constructing wave functions, with spin included, that will satisfy the Pauli principle. Such functions will be described as 'Pauli-compatible'. In extracting an electron density from a given wavefunction, it is now clear that symmetry must play a crucial role.

Given a degenerate set of g exact eigenfunctions of H, the question of how they should be used in constructing an exact wave function, with spin included, was answered in the early days of quantum mechanics. The relevant theorem [14] is well known but is so often ignored as to be worth stating clearly:

> If $\Phi_1, \Phi_2, \ldots \Phi_g$ is a set of g-fold degenerate exact eigenfunctions of the Hamiltonian H, of given energy E, providing an irreducible representation D of the permutation group $_N$, then the normalized function
>
> $$\Psi(\mathbf{x}_1, \mathbf{x}_2, \ldots \mathbf{x}_N) = g^{-\frac{1}{2}} \sum_{\kappa=1}^{g} \Phi_\kappa(\mathbf{r}_1, \mathbf{r}_2, \ldots \mathbf{r}_N) \Theta_\kappa(s_1, s_2, \ldots s_N) \quad (30)$$
>
> will be an exact wavefunction *recognizing the existence of spin and satisfying the Pauli (antisymmetry) principle*, provided the spin eigenfunctions Θ_κ carry the representation, D_S say, *associate*[6] to D. The *only* acceptable space-spin wavefunction of energy E is then the linear combination (30).

The implications of this theorem are far-reaching. If Φ is an exact *non-degenerate* eigenfunction of H ($g = 1$) the only acceptable wavefunction is the product $\Phi\Theta$ in which Θ carries the identity representation, being invariant under all spin permutations, while Φ is totally antisymmetric under permutation of spatial variables. This particular state, however,

---

[6] The 'associate' or 'dual' ($\check{D}$) of a representation D contains matrices $\check{D}(P) = \epsilon_P \tilde{D}(P^{-1})$ where the tilde denotes transposition and $\epsilon_P(= \pm 1)$ is the usual parity factor.

is usually of little interest: it is a state of maximum multiplicity in which all N spins are parallel coupled to a resultant $S = \frac{1}{2}N$. For all other physically allowed states, $g > 1$ and is dependent upon N and S. In fact $g = f_S^N$, where 'Wigner's number' is

$$f_S^N = \frac{(2S+1)N!}{(\frac{1}{2}N+S+1)!(\frac{1}{2}N-S)!}. \tag{31}$$

Thus, for a singlet ground state of a 10-electron system $g = 42$ – this being the number of linearly independent spin eigenfunctions with $N = 10, S = 0$, corresponding to the 42 different paths, in the well known Branching Diagram (BD), that lead from the origin to the point $S = 0$ after N steps. This means that the ground-state singlet wavefunction involves 42 degenerate and linearly independent eigenfunctions of the spin-free Hamiltonian H.

The above considerations are independent of any *orbital approximations* employed in the construction of a wavefunction: they apply to exact solutions of the Schrödinger equation with any given spinless Hamiltonian. In particular, they apply to *model* systems of the kind used in the last Section, where the full Hamiltonian (4) is replaced by

$$\mathsf{H}_0 = \sum_{i=1}^{N} \mathsf{F}(i). \tag{32}$$

Here the 2-electron operators no longer appear (even when F may implicitly include electron interaction terms). But the theorems of DFT should apply equally to both real and model systems; and those of IPM type offer the possibility of conducting general arguments without approximation.

The Hamiltonian (32) provides of course the basis of most *orbital* approximations: it has exact eigenfunctions of the form

$$\Phi^0_{ij\ldots p}(\mathbf{r}_1, \mathbf{r}_2, \ldots \mathbf{r}_N) = \phi_i(\mathbf{r}_1)\phi_j(\mathbf{r}_2)\ldots \phi_p(\mathbf{r}_N), \tag{33}$$

where ϕ_k denotes an eigenfunction of the 1-electron operator F (with eigenvalue ϵ_k, say). The total electronic energy of this independent-particle model (IPM) is

$$E^0_{ij\ldots p} = \epsilon_i + \epsilon_j + \ldots + \epsilon_p \tag{34}$$

and obviously the state is $N!$-fold degenerate, if the 'occupied' orbitals are all different, since there will be $N!$ permuted products $\mathsf{P}\Phi^0_{ij\ldots p}$ giving the same $E^0_{ij\ldots p}$. This permutation degeneracy is resolved by seeking, for a state of given S (which is a *symmetry* label), the f_S^N linearly independent combinations that provide the representation $\check{D}_S$: when spin is admitted and these functions are combined to give the unique antisymmetric space-spin function (30), the permutation degeneracy is removed completely – the resultant wavefunction then being Pauli-compatible. The 'closed-shell' ground state, fundamental in DFT, is obtained by filling the lowest-energy orbitals in (33) *in pairs*, adding spin factors α, β within each pair, and antisymmetrizing to obtain a Slater determinant. But this is merely a convenient way of imposing the correct symmetry without using group theory; and it leads at once to the conclusion that *no spin-orbital can be occupied by more than one electron* – the (limited) statement of the Pauli principle that underpins the whole *aufbau* approach in electronic structure theory. The group theoretical approach is general but less simple. Thus, in the case $N = 10, S = 0$, the irrep D_S is carried by branching diagram functions whose symmetry species may be indicated by Young tableaux (see, for example, Refs.[15][16]) such as

1	3	5	7	9
2	4	6	8	10

– in which electron spin indices appear in the 'boxes'. The tableau is essentially an 'instruction' for generating a symmetry function from a primitive product such as $\alpha\beta\alpha\beta\ldots$, namely, "symmetrize with respect to the indices within each row and antisymmetrize with respect to those in the columns". In this example the result is a Weyl-Rumer function of the form

$$(\alpha(s_1)\beta(s_2) - \beta(s_1)\alpha(s_2)) \times (\alpha(s_3)\beta(s_4) - \beta(s_3)\alpha(s_4)) \times \ldots,$$

as used in valence bond theory. The other linearly independent spin functions are associated with 'standard' tableaux in which the indices appear in the boxes in a different order, but always increasing along the rows and down the columns: in this example there are 42 such possibilities, corresponding to the 42 standard branching diagram functions. The *spatial* functions that carry the associate irrep follow from the spin tableaux by interchanging rows and columns; so the associated orbital function could be obtained by *symmetrizing* with respect to indices (1,2), (3,4), etc. and *antisymmetrizing* with respect to (1,3,5,7,9) and (2,4,6,8,10). On using a primitive product in which the orbitals are identical within pairs (i.e. there is one doubly occupied orbital for each pair) it is easy to show that the sum of products in (30) can be reproduced by antisymmetrizing (in both space and spin variables) a single product

$$\phi_1(\mathbf{r}_1)\phi_1(\mathbf{r}_2)\phi_2(\mathbf{r}_3)\ldots[\alpha(s_1)\beta(s_2) - \beta(s_1)\alpha(s_2)]\ldots$$

– which yields the Slater determinant $|\phi_1\alpha\ \phi_1\beta\phi_2\alpha\ \phi_2\beta....|$.

It is clear from such examples that IPM-type wavefunctions, built up from products of orthogonal spin-orbitals, are most easily handled using Slater determinants. However, it is entirely possible to eliminate the spin factors and to work in terms of spatial functions alone, imposing the correct permutation symmetry by means of a projection operator: this spin-free approach (apart from its practical utility in dealing with non-IPM wavefunctions [17]) is important in revealing the presence of 'hidden' constraints, which must be observed when we pass from the N-electron wavefunction to the electron density and other density functions by integrating over variables that appear to be redundant.

The electron density (1), usually derived from a wave function expressed in terms of Slater determinants, may be obtained more directly in the spin-free formulation, from the set $\{\Phi_\kappa\}$ in (30), without the need for spin integration; and indeed *any one* of the spinless functions Φ_κ may be used for this purpose. But the definition (1) then needs modification: it is replaced by

$$P(\mathbf{r}) = \sum_{i=1}^{N} \int \Phi_\kappa(\mathbf{r}_1, \mathbf{r}_2, ... \mathbf{r}_N)\delta(\mathbf{r} - \mathbf{r}_i)\Phi_\kappa^*(\mathbf{r}_1, \mathbf{r}_2, ... \mathbf{r}_N)d\mathbf{r}_1 d\mathbf{r}_2 ... d\mathbf{r}_N \tag{35}$$

– where each integration may give a different result, owing to the lack of any simple symmetry in the spatial function. A more symmetrical form may be obtained by writing the space-spin Ψ in the form (30) and then using (1): orthogonality of the spin factors leads to the result (normalized)

$$P(\mathbf{r}) = N \int (f_S^N)^{-1} \sum_{\kappa=1}^{f_S^N} \Phi_\kappa(\mathbf{r}, \mathbf{r}_2, ... \mathbf{r}_N)\Phi_\kappa^*(\mathbf{r}, \mathbf{r}_2, ... \mathbf{r}_N) d\mathbf{r}_2 ... d\mathbf{r}_N. \tag{36}$$

This means that spin may be ignored, the electron density being derived in a form *analogous to* (1), viz.

$$P(\mathbf{r}) = N \int \Phi(\mathbf{r}, \mathbf{r}_2, ... \mathbf{r}_N)\Phi^*(\mathbf{r}, \mathbf{r}_2, ... \mathbf{r}_N) d\mathbf{r}_2 ... d\mathbf{r}_N, \tag{37}$$

provided the pure-state density $\Phi\Phi^*$ is replaced by an *ensemble density*:

$$\Phi\Phi^* \longrightarrow (f_S^N)^{-1}(\Phi_1\Phi_1^* + \Phi_2\Phi_2^* + ...\Phi_\kappa\Phi_\kappa^* + ...), \tag{38}$$

in which every degenerate state appears with the same weight. Consequently, when (as in formal DFT) attention is focused solely on the (spatial) electron density, this density cannot be assumed *pure state*

N-representable (i.e. derivable from a single state function of correct symmetry); at best it can be assumed only *ensemble* N-representable.

The implications of these observations do not appear to have been fully explored. It is easily verified explicitly, taking as an example the case of 4 electrons in 4 arbitrary orbitals, that both (35) and (37) – the latter with the ensemble density (38) – yield exactly the same singlet and triplet densities as the conventional approach through (1). But the latter approach appears to be the one universally adopted.

5. Conclusion

As long as N-representability conditions for the pair function Π remain unknown, the goal of rigorously eliminating the wave function in favour of even the 2-DM, let alone the electron density itself, still seems to be unattainable. All that can be done is to impose the proper permutation symmetry conditions on a variational *wave function* and then proceed to the required DMs by integrating over $N-1$ or $N-2$ variables. Even this is not, in general, an easy task – easy as it may be for an IPM 1-determinant wave function.

There are two main aspects of the problem: (i) the *condition* that a spin-free wave function must satisfy in order to provide an acceptable (i.e. Pauli-compatible) starting point; and (ii) how to *impose* the condition on an arbitrary function.

The condition that a given spin-free function Φ belong to irreducible representation $\check{D}_S$ is simply that it be an eigenfunction of the class operator

$$\mathsf{C} = \sum_{i<j=1}^{N} \mathsf{P}_{ij} \tag{39}$$

with eigenvalue

$$c_S^N = N - \tfrac{1}{4}N^2 - S(S+1), \tag{40}$$

as follows readily from the Dirac spin-exchange identity. And this condition involves all the *two*-electron operators P_{ij}.

As for the *imposition* of the condition on a suitable trial function, for a ground state of multiplicity $2S+1$, the standard procedure is to take any N-electron function whatever, Ω say, and apply a projection operator

$$\mathsf{O}_{\kappa\kappa}^{S} = \sum_{\mathsf{P}} \check{D}_S(\mathsf{P})_{\kappa\kappa} \mathsf{P}, \tag{41}$$

for whatever value of the species symbol κ may be convenient, to obtain $\Phi_{S,\kappa} = \mathsf{O}_{\kappa\kappa}^{S}\Omega$. In practice, Ω might be, for example, an orbital product $\Omega_K = \phi_{k_1}\phi_{k_2}\ldots\phi_{k_N}$; and on taking $\kappa = 1$ (for the first tableau of the dual representation $\check{D}_S$) the operator O_{11}^S produces a Weyl-Rumer function of the kind used in spin-free forms of valence bond theory [17]. Unfortunately, the projected function then contains $N!$ terms.

In summary, the construction of a Pauli-compatible variation function from an arbitrary 'root' (of orbital form or not) involves the generation of $N!$ components – and all of them must be used in deriving the 1-DM and the 2-DM. Existing DFT methodology will no doubt continue to give useful results; but the goal of obtaining a rigorous basis for the use of semi-empirical functionals still seems remote.

References

[1] Hohenberg, P., and Kohn, W. (1964). Phys. Rev. **136B**, 864.

[2] Kohn, W., and Sham, L.J. (1965). Phys. Rev. **140A**, 1133.

[3] Kryachko, E.S., and Ludena, E.V. (1990). *Energy Density Functional Theory of Many-Electron Systems*, Kluwer, Dordrecht.

[4] McWeeny, R., (1993) *Methods of Molecular Quantum Mechanics 2ed*, Academic, London. (See Chapter 5)

[5] Weissman, S.I., (1956) J. chem. Phys. **25**, 890; McConnell, H.M. (1958) J. chem. Phys. **28**, 118.

[6] McWeeny, R. (1986) Proc. Ind. Acad. Sci. **96**, 263.

[7] McWeeny, R., and Mizuno, Y. (1961) Proc. Roy. Soc. Lond. **A259**, 554.

[8] Coleman, A.J. (1963) Rev. Mod. Phys. **35**, 668.

[9] Coleman, A.J., and Yukalov, V.I. (2000) *Reduced Density Matrices*, Springer, Berlin.

[10] Davidson, E.R. (1976) *Reduced Density Matrices in Quantum Chemistry*, Academic, New York.

[11] Löwdin, P.-O., (1955) Phys. Rev. **97**, 1474.

[12] McWeeny, R., and Kutzelnigg, W. (1968) Int. J. Quantum Chem. **2**, 187.

[13] McWeeny, R. (1956) Proc. Roy. Soc. Lond. **A235**, 496.

[14] Wigner, E.P. (1959) *Group Theory and its Application to the Quantum Mechanics of Atomic Spectra*, Academic, New York. (Translated from the German edition (1931) by J. J. Griffin.)

[15] Pauncz, R. (1979) *Spin Eigenfunctions: Construction and Use*, Plenum. New York.

[16] McWeeny, R., (1993) *Methods of Molecular Quantum Mechanics 2ed*, Academic, London. (See Chapter 4)

[17] McWeeny, R. (1988) Int. J. of Quantum Chem. **34**, 25; *ibid*, **74**, 87 (1999).

THE NEW FORMULATION OF THE DENSITY FUNCTIONAL THEORY, THE LIMITATION OF ACCURACY OF THE KOHN-SHAM POTENTIAL AND ITS EXPRESSION IN TERMS OF THE EXTERNAL POTENTIAL

Andreas K. Theophilou
"DEMOCRITOS"National Center for Scientific Research, TT 15310 Attica, Greece

Abstract We present some new developments of the new formulation of DFT, where no functional derivatives are used. Further, we point some shortcomings of the standard KS theory, arising from the fact that the Kohn potential (Hartree+exchange and correlation) is expressed in terms of the density. In order to overcome the accuracy limitations due to symmetry requirements, we express the Kohn potential in terms of the external potential. The ground state energy calculation of the present theory in atoms and molecules give deviations from Hartrre-Fock theory ranging from 0.01 to 0.1 per cent. The advanage of the present theory is that it has as input the external potential, which is a known function, whereas in standard DFT theory, one starts with input the electron density about which one has to make an initial guess and after many iterations one gets the correct ground state density.

1. Introduction

Density functional theory (DFT), in the form established by Kohn and coworkers [1, 2] is nowadays a rigorously formulated theory. Problems such as the representability of densities by many particle states have been resolved. In fact, Harriman has shown that every N-particle density is representable by a single Slater determinant [3]. In a formulation by the present author a new functional was introduced the minimum of which gives the Kohn and Shams (KS) state and the ground state energy. The variations needed to get the minimum are in the space of N-particle wave functions and not in the space of densities [4].

Another advantage of this formulation is that the noninteracting states are not restricted to the states of single Slater determinants (SLD), as there are noninteracting states which, because of symmetry requirements, are linear combinations of SLDs. In addition, the problem of functional differentiation in the space of densities is circumvented. We shall give a brief review of this formulation, before going to the problem of symmetry of the KS potential, $V_{KS,}(\mathbf{r})$, which we consider essential, as violation of symmetry may be fatal. Thus we shall prove that all Hartree potentials, $V_H(\mathbf{r})$, and all exchange and correlation potentials, $V_{xc}(\mathbf{r})$, when expressed in terms of the electron density, $\rho(\mathbf{r})$, have limited accuracy, as they do not preserve the symmetry properties of the external potential. For this reason we shall express the *Kohn potential*, $V_K(\mathbf{r}) = V_H(\mathbf{r}) + V_{xc}(r)$ as a mapping of the external potential $V(\mathbf{r})$ in a way that symmetry properties of the external potential are preserved. We give an approximate explicit expression of such a mapping, which was derived by using symmetry and asymptotic conditions for well known systems such as the finite sphere jellium and atoms. Finlly we discuss the accuracy of this mapping by using results derived for the ground state properties of atoms and molecules.

2. The New Formulation of DFT

The Hamiltonian considered here is that of the interacting N-electrons,

$$H = T + H_{int} + \hat{V} \qquad (1)$$

where 'T' is the kinetic energy operator

$$T = \frac{1}{2}\int d^3r \sum_s \nabla\psi^{s\dagger}(\mathbf{r})\nabla\psi^s(\mathbf{r}) \qquad (2)$$

H_{int} is the interaction energy,

$$H_{int} = \frac{1}{2}\int d^3r \int d^3r' \sum_{st} \frac{\psi^{s\dagger}(\mathbf{r})\psi^{t\dagger}(\mathbf{r}')\psi^t(\mathbf{r}')\psi^s(\mathbf{r})}{|\mathbf{r}-\mathbf{r}'|}, \qquad (3)$$

$$\hat{V} = \int d^3r \hat{\rho}(\mathbf{r})V(\mathbf{r}) \qquad (4)$$

$\hat{\rho}(\mathbf{r})$ is the density operator

$$\hat{\rho}(\mathbf{r}) = \sum_s \psi^{s\dagger}(\mathbf{r})\psi^s(\mathbf{r}) \qquad (5)$$

and $\psi^{s\dagger}(\mathbf{r})$, $\psi^s(\mathbf{r})$ are the fermion field creation and annihilation operators. The summation over s and t indicates summation with respect to the spin indices.

We next define the basic functionals for the formulation of DFT, and establish a variational principle with variations in the space of the N-particle states.

The following functionals are defined for any state $|\Psi\rangle \in H$ where H is the Hilbert space of the N-particle wave functions

$$K(\Psi) = \inf\{\langle\Phi|T|\Phi\rangle : \langle\Phi|\hat{\rho}(\mathbf{r})|\Phi\rangle = \langle\Psi|\hat{\rho}(\mathbf{r})|\Psi\rangle\} \quad (6)$$

This functional defines a kinetic energy for every state $|\Psi\rangle \in H$ which is not its own. $K(\Psi)$ is the minimum kinetic energy in the space of all states $|\Phi\rangle$ which have the same density as $|\Psi\rangle$. One can prove that this infimum is a minimum, i.e. there is a state $|\Phi\rangle \in H$ such that $K(\Psi) = \langle\Phi|T|\Phi\rangle$ [40, 6]., In a similar way we define the functional

$$L(\Psi) = \inf\{\langle\Phi|T + H_{int}|\Phi\rangle : \langle\Phi|\hat{\rho}(\mathbf{r})|\Phi\rangle = \langle\Psi|\hat{\rho}(\mathbf{r})|\Psi\rangle\} \quad (7)$$

where again the infimum is a minimum [40, 6] . Note that these functionals are related to the ones defined by Levy [7], the difference being that our functionals are not defined over the space of densities, but over all states of the Hilbert space. Finally we define the somewhat strange functional $G(\Psi; V)$ which when minimized will give the KS states having the same density as the exact ground state.

$$G(\Psi; V) = L(\Psi) + \langle\Psi|T|\Psi\rangle - K(\Psi) + \int d^3r \langle\Psi|\hat{\rho}(\mathbf{r})|\Psi\rangle V(\mathbf{r}) \quad (8)$$

Note that all these functional are defined for the whole Hilbert and not in the space of densities. Thus the variations should be in the Hilnert space of states.

Since $\langle\Psi|T|\Psi\rangle - K(\Psi) \geq 0$ one concludes that the minimum of $G(\Psi, V) = G(\Phi)$ is obtained when equality holds and also that

$$Inf G(\Psi) = \langle\Psi_0|H|\Psi_0\rangle = E_0 \quad (9)$$

where $|\Psi_0\rangle$ is the exact ground state of the interacting system [4]. In order to derive the KS equation

$$T|\Phi\rangle + V_{KS}|\Phi\rangle = E|\Phi\rangle, \quad (10)$$

we shall use the hypothesis, that for every $|\Phi\rangle$, which minimizes $K(\Psi)$, there is a potential $V_{KS}(r, V)$, for which the above equation is satisfied or in other words, for each interacting ground state, there is a noninteracting one which obeys the KS equation. We prove this hypothesis at the end of this chapter (Theorem 2). Note that the above equation does not exclude an excited state from being a minimum kinetic energy state. In fact Levy has shown that excited states with minimum kinetic energy exist.

In standard DFT, one writes $V_{KS}(\mathbf{r}) = V(\mathbf{r}) + V_K(\mathbf{r}, \rho)$, where the Kohn potential is expressed as a Hartree $V_H(\mathbf{r}, \rho)$, and an exchange and correlation potential $V_{xc}(\mathbf{r}, \rho)$,

$$V_K(\mathbf{r}, \rho) = V_H(\mathbf{r}, \rho) + V_{xc}(\mathbf{r}, \rho), \tag{11}$$

where

$$V_H(\mathbf{r}, \rho) = \int d^3r' \frac{\rho(\mathbf{r}')}{|\mathbf{r} - \mathbf{r}'|} \tag{12}$$

The advantage of expressing the KS potential in terms of the density is that we know exactly the functional form of $V_H(\mathbf{r}, \rho)$ and to some approximation $V_{xc}(\mathbf{r}, \rho)$, as there is a lot of experience from the Hartree-Fock approximation, although the two exchange potentials refer to different concepts. We shall not use the above separation of the Kohn potential and also we shall not express it in terms of the density but in terms of the external potential V. The reason is that it was possible for us to find universal symmetry preserving mappings of V onto V_K whereas this is not an easy task when expressed in terms of the density ρ. As one can see from the expression of $V_H(\mathbf{r}, \rho)$ above any asymmetry of the density is transferred to the left hand side and in general the density does not have the symmetry of the external potential, whereas the KS potential should have the same symmetry as V.

The above formulation of DFT is useful not only because no functional derivatives are use but also for the reason that we do not confine the space of KS states in the space of single Slater determinants. This is necesary, as in some cases, because of symmetry requirements, it is not possible to express a noninteracting state as a single Slater determinant. Such an example is a ^1P two-particle state with $S = 0$ which has the following form

$$|\Phi\rangle = \frac{1}{\sqrt{2}}(a_0^{\uparrow +} a_1^{\downarrow +} + a_1^{\uparrow +} a_0^{1\downarrow +})|0\rangle \tag{13}$$

The above formulation is also possible for the subspace theory of excited states and it is applicable for the case of degeneracy, where there is no one to one correspondence between potential and density [9].

For the existence of the KS potential, we need to show that the set of states satisfying the density constraint in the neighborhood of the minimizing state, with vanishing first order variations of the kinetic energy, is not empty, i.e. that

Lemma 1. If $|\Phi\rangle$ is a minimizing state for $K(\Phi)$, then in an ε−neighbourhood of $|\Phi\rangle$, there is a state $|\Phi_\varepsilon\rangle$, having the same density such that $T(\Phi_\varepsilon) - T(\Phi) \langle c^2\varepsilon^2$, for $|\Phi_\varepsilon - \Phi|\langle\varepsilon$. The proof is restricted to zero current minimizing states in order to avoid unnecessary complications..

Proof: The state $|\Phi_\varepsilon\rangle$ resulting by substituting $\Phi(\mathbf{r}_1, \mathbf{r}_2, \ldots, \mathbf{r}_N)$ by $\Phi_\varepsilon(\mathbf{r}_1, \mathbf{r}_2, \ldots, \mathbf{r}_N) = e^{i\varepsilon \sum u(\mathbf{r}_i)} \Phi(\mathbf{r}_1, \mathbf{r}_2, \ldots, \mathbf{r}_N)$, has the same density as $|\Phi\rangle$. Note that with this substitution every Slater determinant is transformed into a Slater determinant as the KS orbitals $\phi_i(\mathbf{r})$ are transformed to $\phi_i^\varepsilon(\mathbf{r}) = e^{i\varepsilon u(\mathbf{r})} \phi_i(\mathbf{r})$, which are mutually orthogonal. Since $|\Phi\rangle$ corresponds to a real wave function, all $\phi_i(\mathbf{r})$ can be chosen real. Then it can be easily deduced by direct calculation that $\langle \Phi^\varepsilon | T | \Phi^\varepsilon \rangle = \langle \Phi | T | \Phi \rangle + \varepsilon^2 \int d^3r \nabla u(\mathbf{r}) \cdot \nabla u(\mathbf{r}) \rho(\mathbf{r})$. We choose $u(\mathbf{r})$ to belong to H^1, the space of square integrable functions of finite kinetic energy and to have a finite $|\nabla u(\mathbf{r})|$. By using the fact that $\rho(\mathbf{r}) \geq N/\Omega_0$ only in a finite volume $\Omega \langle \Omega_0$, because of the inequalities $N\Omega/\Omega_0 = \int_\Omega d^3r N/\Omega_0 \langle \int_\Omega d^3r \rho(\mathbf{r}) < N$, we conclude that $\int_\Omega d^3r \nabla u(\mathbf{r}) \cdot \nabla u(\mathbf{r}) \rho(\mathbf{r})$ is finite. Then in the volume Ω_c complementary to Ω we have the inequality

$$\int_{\Omega c} d^3r \nabla u(\mathbf{r}) \cdot \nabla u(\mathbf{r}) \rho(\mathbf{r}) < \int_{\Omega c} d^3r \nabla u(\mathbf{r}) \cdot \nabla u(\mathbf{r}) N/\Omega_0.$$

This is finite too since $u(\mathbf{r})$ belongs to H^1. Then it follows that $\int d^3r \nabla u(\mathbf{r}) \cdot \nabla u(\mathbf{r}) \rho(\mathbf{r}) = c^2 < \infty$. Therefore, $T(\Phi_\varepsilon) \langle T(\Phi) + \varepsilon^2 c^2$. Thus $T(\Phi_\varepsilon) - T(\Phi) < c\varepsilon^2$, Q.E.D.

This lemma implies that for any state $|\Phi\rangle$ there are states $|\Phi_\varepsilon\rangle$ with the same density such that $\lim_{\varepsilon \to 0}(\frac{T(\Phi_\varepsilon) - T(\Phi)}{\varepsilon}) = 0$. Note that with this lemma we did not show that all variations compatible with the density constraint give this limit.

In order to derive the KS equation it is necessary to elaborate more into the spaces of the single state representable densities by looking into the Banach spaces in which the physical densities can be represented. For this matter we shall show that

Lemma 2 Each single state representable density, $\rho(\mathbf{r})$, corresponds to a function $f(\mathbf{r}) = \sqrt{\rho(\mathbf{r})}$ which belongs to H^1.

We give the proof by using field theory as it is short. A more elementary proof can be given in terms of the natural orbitals. This in fact is Theorem 1 in Ref [40] expressed in physical terms.

Proof: We note first that $\nabla f(\mathbf{r}) = \frac{\nabla \rho(\mathbf{r})}{2\sqrt{\rho(\mathbf{r})}}$, and therefore $|\nabla f(\mathbf{r})|^2 = \frac{(\nabla \rho(\mathbf{r}))^2}{4\rho(\mathbf{r})}$. Since the state representable densities are of the form $\rho(\mathbf{r}) = \langle \Phi | \psi^+(\mathbf{r}) \psi(\mathbf{r}) | \Phi \rangle$ it follows that

$$|\nabla\rho(\mathbf{r})| = |\langle\Phi|\nabla\psi^+(\mathbf{r})\psi(\mathbf{r}) + \psi^+(\mathbf{r})\nabla\psi(\mathbf{r})|\Phi\rangle|$$
$$\leq |\langle\Phi|\nabla\psi^+(\mathbf{r})\psi(\mathbf{r})|\Phi\rangle| + |\langle\Phi|\psi^+(\mathbf{r})\nabla\psi(\mathbf{r})|\Phi\rangle \quad (15)$$

By taking into account the equality $\langle\Phi|\nabla\psi^+(\mathbf{r})\psi(\mathbf{r})|\Phi\rangle = \langle\nabla\psi(\mathbf{r})\Phi|\psi(\mathbf{r})|\Phi\rangle$ and using the Schwarz inequality we have

$$|\langle\nabla\psi(\mathbf{r})\Phi|\psi(\mathbf{r})|\Phi\rangle| \leq \langle\Phi|\nabla\psi^+(\mathbf{r})\nabla\psi(\mathbf{r})|\Phi\rangle^{1/2}\langle\Phi|\psi^+(\mathbf{r})\psi(\mathbf{r})|\Phi\rangle^{1/2} \quad (16)$$

Similarly

$$|\langle\psi(\mathbf{r})\Phi|\nabla\psi(\mathbf{r})|\Phi\rangle| \leq \langle\Phi|\nabla\psi^+(\mathbf{r})\nabla\psi(\mathbf{r})|\Phi\rangle^{1/2}\langle\Phi|\psi^+(\mathbf{r})\psi(\mathbf{r})|\Phi\rangle^{1/2} \quad (17)$$

After adding, we find

$$|\nabla\rho(\mathbf{r})| \leq \langle 2\Phi|\nabla\psi^+(\mathbf{r})\nabla\psi(\mathbf{r})|\Phi\rangle^{1/2}\rho^{1/2}(\mathbf{r}) \quad (18)$$

and therefore

$$\frac{(\nabla\rho(\mathbf{r}))^2}{4\rho(\mathbf{r})} \leq \langle\Phi|\nabla\psi^+(\mathbf{r})\nabla\psi(\mathbf{r})|\Phi\rangle \quad (19)$$

The left hand side is $(\nabla f(\mathbf{r}))^2$. After integrating, the right hand side become the kinetic energy times 2. So finally one gets the inequality

$$\int d^3r (\nabla f(\mathbf{r}))^2 \leq 2\langle\Phi|T|\Phi\rangle \quad (20)$$

Since our states $|\Phi\rangle$ have finite kinetic energy as by definition belong to H^1 it follows that $\nabla f \in L^2$ and since $\int d^3r f^2(\mathbf{r})) = \int d^3r \rho(\mathbf{r}) = N$ it follows that $f = \rho^{1/2}(\mathbf{r}) \in H^1$, q.e.d

The question is wether the set of single state representable functions is a subset of H^1 or the two spaces coincide. Harriman's was able to construct an N-particle wave function $|\Phi\rangle$ for a given density $\rho^{1/2}(\mathbf{r})$. If we now confine the space of densities to the H^1 space, we find by direct calculation that the kinetic energy of these N-particle states is finite. Thus the two spaces coincides i.e.

Theorem 1. The space of representable densities is the H^1 space.

One can now use the above inequality together with the Sobolev inequality

$$3(\pi/2)^{4/3}[\int d^3r |f(\mathbf{r})|^6]^{1/3} \leq \int d^3r (\nabla f(\mathbf{r}))^2 \quad (21)$$

to prove that the single state representable density ρ satisfies the inequality

$$3(\pi/2)^{4/3}[\int d^3r|\rho(\mathbf{r})|^3]^{1/3} \leq 2\langle\Phi|T|\Phi\rangle \qquad (22)$$

i.e. ρ has a finite norm in L^3. Thus we conclude

Lemma 3: All single state representable densities belong to the Banach space with norm $\parallel \rho \parallel_3 = [\int d^3r|\rho(\mathbf{r})|^3]^{1/3}$.

Since representable densities ρ belong to L^3 and $\nabla\sqrt{\rho} \in L^2$, one can show that $\rho \in L^p$ for all $p \in [1,3]$ [40]. It is to be reminded that by L^p we denote the space for which $\int d^3r|\rho(\mathbf{r})|^p]^{1/p} \leq \infty$. This property is useful in order to define the space of the potentials, which is the dual of the space of densities. By taking into account that all continuous linear functionals in this space can be represented by the mapping $v(\rho) = \int d^3rV(\mathbf{r})\rho(\mathbf{r})1001$[8], we conclude that the space of $V(\mathbf{r})$, in which this integral is well defined is the dual to the space of densities. This is the space $V = L^{\frac{3}{2}} + V_\infty$, [40] where V_∞ is the space of bounded functions $V(\mathbf{r})$. But we know that by a theorem of functional analysis the dual of a normed space is a Banach space and therefore it is a complete and separable space [8]. We can use then a complete set of linearly independent potentials V_k to determine uniquely a density from the coefficients $c_k = v_k(\rho) = \int d^3rV_k(\mathbf{r})\rho(\mathbf{r})$, i.e. if the coefficients of two densities coincide then the densities are considered identical[1].

Let us search now for the minimum of $T(\Phi) = \langle\Phi|T|\Phi\rangle$ under the constraint : $\langle\Phi|\hat{\rho}(\mathbf{r})|\Phi\rangle = \rho(\mathbf{r})$ where $\rho(\mathbf{r})$, is a single state representable density. By the completeness theorem we can substitute the density constraint by

$$\int d^3rV_k(\mathbf{r})\langle\Phi|\hat{\rho}(\mathbf{r})|\Phi\rangle = v_k(\rho), \; k=1,2... \qquad (23)$$

---

[1] In order to have a complete system of functions the space must be complete and separable. Although there are some spaces not having a countable basis although they have this property we consider that these are very strange spaces to be acceptble by physicists. The problem is that the norm of these spaces is defined by integration in a finite volume whereas our wavefunctions and densiites extend to infinity. Thus strictly speaking, the complete system of functions does not exist neither for the potentials nor for the wavefunctions. This is not a serious constraint for a physical system as the potential described by the Hamiltonian in a real physical system is of finite rannge. Thus, here we use the concept of completeness of potentials in the same way that is currently used for wavefunctions. However in Physics usually we make the assumption that there is a physical subspace of the complete space which is separable. In this way we use the expansion in terms of a complete set of square integrable functions. The hypothesis that our physical densities and potentials form a separable space is made here.
It is to be reminded that a metric space is complete if every Cauchy sequence converges in the space. A space S is called separable if it has a countable subset which is dense in it.

Let $|\Phi\rangle$ the minimizing state, and $|\Phi_\varepsilon\rangle = |\Phi\rangle + \varepsilon|\Theta\rangle$ a state in the neighborhood of $|\Phi\rangle$ compatible with the density constraint. Then, after introducing the expression of $|\Phi_\varepsilon\rangle$ into the equation

$$Lim_{\varepsilon \to 0} \int d^3r V_k(\mathbf{r}) \frac{\langle \Phi_\varepsilon|\hat{\rho}(\mathbf{r})|\Phi_\varepsilon\rangle - \langle \Phi|\hat{\rho}(\mathbf{r})|\Phi\rangle}{\varepsilon} = 0, \; k = 1, 2... \quad (24)$$

it follows that

$$\langle \Theta|\hat{V}_k|\Phi\rangle = \int d^3r V_k(\mathbf{r}) \langle \Theta|\hat{\rho}(\mathbf{r})|\Phi\rangle = 0 \quad (25)$$

Thus the space U of the states $|\Theta\rangle$ along which variations are allowed is normal to the space $U_\perp$ spanned by the states $V_k|\Phi\rangle k = 1.2...$ and $|\Phi\rangle$, itself. We now prove

Lemma 4. All first order variations of $T(\Phi)$ compatible with the density constraints satisfy the equation

$$\lim_{\varepsilon \to 0} \left(\frac{T(\Phi_\varepsilon) - T(\Phi)}{\varepsilon} \right) = 0 \quad (26)$$

Proof: By using the explicit form of $|\Phi\varepsilon\rangle$ we find $T(\Phi_\varepsilon) - T(\Phi) = \varepsilon(\langle \Theta|T|\Phi\rangle + \langle \Phi|T|\Theta\rangle) + \varepsilon^2 \langle \Theta|T|\Theta\rangle \rangle 0$. After dividing by ε and taking the limit for $\varepsilon \to 0$ we find

$$\lim_{\varepsilon \to 0} \left(\frac{T(\Phi_\varepsilon) - T(\Phi)}{\varepsilon} \right) = \langle \Theta|T|\Phi\rangle + \langle \Phi|T|\Theta\rangle \geq 0 \quad (27)$$

Consider now the state $|\Phi'_\varepsilon\rangle = |\Phi\rangle - \varepsilon|\Theta\rangle$. It is straightforward to verify that this state satisfies also the density constraints. Using again the fact that $|\Phi\rangle$ is the minimizing state we find in the same way

$$\lim_{\varepsilon \to 0} \left(\frac{T(\Phi'_\varepsilon) - T(\Phi)}{\varepsilon} \right) = -(\langle \Theta|T|\Phi\rangle + \langle \Phi|T|\Theta\rangle) \geq 0 \quad (28)$$

and from these two inequalities it follows that $\langle \Theta|T|\Phi\rangle + \langle \Phi|T|\Theta\rangle = 0$. By repeating the procedure using the state $|\Phi''_\varepsilon\rangle = |\Phi\rangle - i\varepsilon|\Theta\rangle$ we find

$$\langle \Theta|T|\Phi\rangle = 0 \quad (29)$$

This proves our lemma.

¿From the above it follows that the vector $T|\Phi\rangle$ is normal to the space of acceptable variations U and therefore it has components only in space $U_\perp$ spanned by the states $V_k|\Phi\rangle$ and $|\Phi\rangle$ itself. Thus $T|\Phi\rangle$ is a linear combination of the states $V_k|\Phi\rangle$ and $|\Phi\rangle$ itself i.e. $T|\Phi\rangle = -\sum_k \lambda_k \hat{V}_k |\Phi\rangle + E|\Phi\rangle$. Thus we have

Theorem 2. The state which minimizes the kinetic energy satisfies the eigenvalue equation

$$T|\Phi\rangle + \hat{V}|\Phi\rangle = E|\Phi\rangle \tag{30}$$

with

$$\hat{V} = \sum_k \lambda_k \hat{V}_k \tag{31}$$

The so obtained $|\Psi\rangle$ depends on the parameters λ_k. By inserting this in the constraint equation we get an algebraic equation for the parameters λ_k. For bound states we demand our potentials be real. However, if real parameters λ_k do not exist, then there is no solution. Also the parameters may specify an extremum and not necessarily a minimum. Conversely if we restrict ourselves to only some constraints, e.g. for k=1 to 5, then these constraints will give an eigenvalue equation which does not imply that the extremum associated with these constraints is a minimum under the density constraint. However in case the densities correspond to minima under a finite number of density constraints

$$\langle u|\hat{V}_k|\Phi\rangle = \int d^3 r V_k(\mathbf{r})\langle u|\hat{\rho}(\mathbf{r})|\Phi\rangle = 0, \ k = 1, 2, ...M \tag{32}$$

it is possible to correspond to different ground state densities. The possible λ_k are determined after solving the eigenvalue equation and substituting the solutions into the constraint equations. Then we may find different groups of λ_k and to each such group corresponds a potential $\hat{V} = \sum_k \lambda_k \hat{V}_k$. One of these solutions corresponds to minimum kinetic energy state. Concerning the KS equation, the above solution may be a ground state or an excited state for the given potential. Although, this equation was derived by applying the minimum kinetic energy variational principle, there is no way to check beforehand, whether for the given potential we have a minimum or an extremum of the kinetic energy for given potential, i.e by determining the Lagrance multipliers.

In order to elucidate the situation let us take the simplest case of a single constraint, i.e. the condition $\int d^3 r V(\mathbf{r}) \rho(\mathbf{r}) = c$. Then, the resulting equation after applying the minimum principle for the kinetic energy is

$$T|\Phi\rangle + \lambda \hat{V}|\Phi\rangle = E|\Phi\rangle \tag{33}$$

and as $|\Phi\rangle$ is not a linear function of the parameter λ, the condition may give more than one real solutions for λ.

It is time to make clear that the mathematicians define functional derivatives in a very different way [8]. Thus, if $K(\varphi)$ is a functional in a space S, then the mathematical functional derivative is defined as a mapping A which depend on φ but not on u, for which the following relation holds

$$\lim_{\varepsilon \to 0} \frac{K(\varphi + \varepsilon u) - K(\varphi)}{\varepsilon} = A(\varphi)u \qquad (34)$$

Then one can show that the operator A **is not** a local function of the position variable. For some cases A is related to a local function, but itself in general is an operator. Thus let us take the case that

$$K(\varphi) = \int g(\varphi(x), x) dx \qquad (35)$$

Then using the above definition we have

$$\frac{1}{\varepsilon}\int g(\varphi(x)+\varepsilon u(x),x) - g(\varphi(x),x)dx = \frac{1}{\varepsilon}\int \varepsilon u(x)\frac{d}{d\varphi}g(\varphi(x),x)$$
$$= \int \frac{d}{d\varphi}g(\varphi(x),x)u(x) \quad (36)$$

Then, in this case, what we call in physics functional derivative is the function $\frac{d}{d\varphi}g(\varphi(x),x)$. We can refer to this function as the 'local derivative' or DFT derivative. We note here that other functional derivatives, e.g. depending also on the variation u can be defined [8].

2.1 Asymmetries

Symmetries in physics are very important not only for making the solution of a problem easier but also for specifying many of the qualitative features of a solution. In an earlier paper we proved

Theorem 3. (Theophilou and Papaconstantinou [10]): The KS orbitals, in a KS potential not having the symmetry of the external potential, do not belong to a definite Irrep of the symmetry group of the Hamiltonian.

We shall illustrate the content of the above theorem by an example, assuming that an eigenfunction of the KS equation exists which belongs to the identity Irrep of the group G, under which the external potential V is invariant while the KS V_{KS} potential does not have this symmetry. Then by expressing V_{KS} in terms of its Irreducible components relative to G, we have

$$V_{KS}(\mathbf{r}) = V_{KS}^0(\mathbf{r}) + \sum_{\Gamma \neq 0} V_{KS}^\Gamma(\mathbf{r}) \qquad (37)$$

Assume now that a solution belonging to the identity Irrep of G exists. Then the KS equation is

$$-\frac{1}{2}\nabla^2\phi^0(\mathbf{r}) + (V_{KS}^0(\mathbf{r}) + \sum V_{KS}^\Gamma(\mathbf{r}))\phi^0(\mathbf{r}) = \varepsilon\phi^0(\mathbf{r}) \qquad (38)$$

As the operator ∇^2 is invariant under I_3^3, the group of all rotations and translations, it is also invariant under any of its subgroups, like e.g. G. Then $\nabla^2\phi_\gamma^\Gamma(\mathbf{r}) = \psi_\gamma^\Gamma(\mathbf{r})$, i.e. the kinetic energy operator maps a function belonging to certain Irrep Γ to another one belonging to the same Irrep. Since functions belonging to different Irreps are orthogonal and therefore linearly independent, it follows from the above equation that $V_{KS}^\Gamma(\mathbf{r})\phi^0(\mathbf{r}) = 0$ and therefore $V_{KS}^\Gamma(\mathbf{r}) = \mathbf{0}$ when Γ is not the identity Irrep of G.

The same symmetry requirement holds for the many particle KS wave function. The proof is as in the single-particle case.

Theorem 4. A necessary condition that the non-interacting many-particle KS state should belong to the same Irreps as the exact interacting state, is that the KS potential must be invariant under the symmetry group of the external potential.

The above theorems put a strong condition on the expressions of the Kohn potential $V_K(\mathbf{r}, \rho)$. In general the density does not belong to the identity Irrep of the symmetry group of the Hamiltonian and therefore it is not easy to find mappings of the density to V_K which satisfy the above condition. For the case of exact symmetries one can express V_K in terms of $\rho^0(\mathbf{r})$, the component of $\rho(\mathbf{r})$ belonging to the identity Irrep of the symmetry group of the external potential, which is equal to the subspace density. However in the case of a small symmetry breaking, which makes the ground state nondegenerate, this is not possible. In this case all existing $V_K(\mathbf{r}, \rho)$ do not comply with this condition and therefore their accuracy is limited. Thus e.g. in the expression giving the Hartree potential any asymmetry on the density is transmitted to the left hand side. The exchange and correlation potential even in its simplest form which is $V_{xc}(\mathbf{r}; \rho) = c\rho^{1/3}(r)$ does not comply with the symmetry constraint either. Details about the functional forms $V_{xc}(\mathbf{r}; \rho)$ can be found in standard books of DFT [11]. We also cite some recent advances on this topic in Ref [12]

3. The KS potential as a mapping of the external potential.

In order to avoid the inconsistencies related to symmetry, we tried to find expression of the Kohn potential in terms of the density, which do not separate the it into a Hartree and a V_{xc} part. But we were successful only in the case of special symmetries. For this reason we decided to express the KS potential in terms of the external potential and the number of particles [14]. After rejecting many mappings, because they did not obey the proper asymptotic conditions for atoms and the finite radius jelium, we adopted the following mapping as the simplest one satisfying these conditions

$$V_{KS}(\mathbf{r},V) = \int d^3r f(|\mathbf{r}'-\mathbf{r}|)\nabla^2 V(\mathbf{r}')$$ (39)

The advantage of this expression, beyond the preservation of symmetry properties is that a constant potential does not have any effect on the exchange and correlation potential. By using the asymptotic relations for atoms and the finite radius jellium we found that a proper expression for $f(r)$ is

$$f(r) = -\frac{1}{r}\left(\frac{Q}{N} + (Q - c - 1)e^{-r/a}\right)$$ (40)

where N is the number of electrons and

$$Q = \int d^3r \nabla^2 V(\mathbf{r})$$ (41)

In practice we considered also that Q is a parameter. One could use this potential to calculate the self-consistent KS state without resuming to iterations as one does with standard DFT. Then the total energy of a many particle system could be calculated by using one of the many $E_{xc}(\rho)$ functionals for the exchange and correlation energy. However, as we do not know the exact values of the parameters a and c we decided to apply this expression using the optimized effective potential theory (OPT) [13], which recently has been developed to include excited state calculations including multiplet structure with considerable success by A. Nagy [16]. We calculated the ground state energies for 20 neutral atoms. We found that the deviations from the exact energies for $Z = 2$ to 12 were smaller than 0.05% while those from $Z = 13$ to 21 ranged from 0.12 to 0.2%. The values of a ranged from 0.38 to 0.67 and those of c from 0 to 1.1. Our calculations for $H_2, LiH, and BH$ molecules for a wide range of internuclear distances give deviations ranging from 0.01 to 0.1 per cent. Details will be published elsewhere.

Thus the present method looks promising. Obviously one has to try the present functional form to more complex systems and make improvements which obviously will involve nonlinear mappings.

Acknowledgments: I would like to thank Professor W. Kohn for drawing my attention on the problem of small symmetry breaking perturbations in relation to DFT and endless discussions with him. The author would like to thank Prof M. Levy and Dr N. Gidopoulos for constructive discussions. He would also like to thank Prof V.N. Glushkov and Dr P.G. Papaconstantinou both for discussions and for communicating their results on molecules and atoms.

References

[1] W. Kohn, Phys. Rev. A **34** 737 (1986)P. Hohenberg and W. Kohn, Phys. Rev. B **136**, 864 (1964).

[2] W. Kohn and L.J. Sham, Phys. Rev. A **140,** 1133 (1965); E.K.U.Gross, L.N.Oliveira and W.Kohn, Phys. Rev. A **37**, 2805 (1988); *Ibid* **37** 2809 (1988), *Ibid* **37** 2821 (1988).

[3] Harrimann. Phys. Rev. A **24,** 680 (1965)

[4] A.K. Theophilou, Int. J. Quant. Chem, **69,** 461 (1998).

[5] E. Lieb, Int.J.Quant. Chem. **24** 243 (1983).

[6] N. Hadjisavvas and A.K. Theophilou, Phys. Rev. A **30**, 2002 **(1984)** ; Ibid A **32**, 720 **(1985).**

[7] M.Levy, Proc. Nat. Acad. Sci. USA, **76**, 6062 (1979).

[8] R.D. Milne, Applied functional analysis, theorem 3.5.1, p155, Pitman Publishing Inc, Massachusettes, USA, 1980.

[9] A.K. Theophilou and P. Papaconstantinou, Phys. Rev A, **61**, 022502 (2000): A.K. Theophilou, J.Phys.C**12**,5419 (1979).

[10] A.K. Theophilou and P. Papaconstantinou, J. Molec. Structure (Theochem), **501-502**, 85 (2000); A.K. Theophilou in *Symmetry and Structural Properties of Condensed Matter Physics*, T. Lulek, W. Florek and B. Lulek Editors, World Scientific, Singapore, London (1996); Int. J. Quant. Chem, **61**, 333 (1997) p.125; *The single particle-density in Physics and Chemistry*, N.H. March and W. Deb, Eds. ;

[11] R.G. Parr and W. Yang, *Density Functional Theory in Atoms and Molecules* (Oxford University Press, New York, 1989); R.M. Dreizler and E.K.U. Gross, *Density Functional Theory (Springer, Berlin, 1990);* N.H. March and W. Deb, Eds. *The Single Particle-density in Physics and Chemistry,* (Academic Press, London, 1987); R.F. Nalewajsky, *Density Functional Theory* (Springer Series: Topics in Current Chemistry, Springer-Verlag, Berlin, Heidelberg 1996);

[12] .N. Gidopoulos, P.G. Papaconstantinou and E.K.U. Gross, Phys. Rev. Lett. **88**, 033003 (2002); M. Higuchi and K. Higuchi, Phys. Rev. B **65**, 195122 (2002); P.P. Rushton, D.J. Tozer and S.J. Clark, Phys. Rev. B **65**, 193106 (2002); M. Fuchs and X. Gonze, Phys. Rev. B **65**, 235109 (2002); M. Nekkovee, W.M.C. Foulkes and R.J. Needs, Phys. Rev. Lett. **87**, 036401 (2001);A.F. Bonetti, E. Engel, R.N. Schmid and R.M. Dreizler, Phys. Rev. Lett. **86**, 2241 (2001); Q.V. Gritsenko and E.J. Baerends, Phys. Rev. A **64**, 042506 (2001); L.C. Balbas, J.L. Martins and J.M. Soler, B **64**, 165110 (2001); T. Tsuneda and K. Hirao, Phys. Rev. B **62**, 15527 (2000); J.P. Perdew, A.Zupan and P. Blaha, Phys. Rev. Lett. **82**, 2544036401 (2001); A. Goerling, Phys. Rev. Lett. **83**, 5459 (1999).

[13] J. D. Talman and W. F. Shadwick, Phys. Rev. A **14**, (1976) 36; Jiqiang Chen, J. B. Krieger,Yan Li and G.J. Iafrate, Phys. Rev. A **54**, (1996) 3939; J. B. Krieger,

Y. Li and G. J. Iafrate, Phys. Lett. A **146** (1990) 256; Phys. Rev. A **45** (1992) 101; Phys. Rev. A **46** (1992) 5453; Int. J. Quantum. Chem. **41** (1992) 489

[14] A.K. Theophilou and P. Papaconstantinou in *Symmetry and Structural Properties of Condensed Matter Physics*, T. Lulek, B. Lulek and A.Wal, Editors, World Scientific, Singapore, London (2000) pp 196-204).

[15] C.R. Colle and D. Salvetti, Theor. Chim. Acta **37**, 329 (1975);

[16] Á. Nagy, J. Phys. B **32** (1999) 2841; Int. J. Quantum. Chem. S. **29** (1995) 297; Phys. Rev. A **57** (1998) 1672; Phys. Rev. A **55** (1997) 3465; J. Phys. B **32** (1999) 2841; Int. J. Quantum. Chem. S. **29** (1995) 297; Nagy, Phys. Rev. A **57** (1998) 1672; Á. Nagy, Phys. Rev. A **55** (1997) 3465.

FUNCTIONAL N-REPRESENTABILITY IN DENSITY MATRIX AND DENSITY FUNCTIONAL THEORY

E.V. Ludeña
Centro de Química, Instituto Venezolano de Investigaciones Científicas, IVIC
Apartado 21827, Caracas 1020-A, Venezuela
eludena@ivic.ve

V.V. Karasiev
Centro de Química, Instituto Venezolano de Investigaciones Científicas, IVIC
Apartado 21827, Caracas 1020-A, Venezuela
vkarasev@ivic.ve

P. Nieto
Centro de Química, Instituto Venezolano de Investigaciones Científicas, IVIC
Apartado 21827, Caracas 1020-A, Venezuela
panf@telcel.net.ve

Abstract   We discuss the problem of functional N-representability both in density-matrix theory and density functional theory. For the case of the 1-matrix, we show that functional N-representability imposes conditions on the reduced single-particle operator as well as on the 1-matrix. We illustrate this point by applying a constructive method in order to obtain a functionally pure-state N-representable single-particle operator acting on an ensemble N-representable 1-matrix. For the case of density functional theory, we deal with functional N-representability in terms of local scaling transformations. An exact functional for Hooke's atom generated using these transformations is presented.

Keywords: reduced matrix theory, density functional theory, N-representability

1. Introduction

In view of the fact that the time-independent Schrödinger equation $\widehat{H}_v^N \Psi_k^{v,N} = E_k^{v,N} \Psi_k^{v,N}$ (where we emphasize that the eigenfunctions depend both on the external potential v and on the total number of electrons N) corresponding to the N-particle Hamiltonian

$$\begin{aligned}\widehat{H}_v^N &= \sum_{i=1}^{N} -\frac{1}{2}\nabla_{\mathbf{r}_i}^2 + \sum_{i=1}^{N-1}\sum_{j=i+1}^{N} \frac{1}{|\mathbf{r}_i - \mathbf{r}_j|} + \sum_{i=1}^{N} v(\mathbf{r}_i) \\ &= \widehat{H}_0^N + \widehat{V}_{ext}^N \end{aligned} \quad (1)$$

cannot be solved analytically except for very few and simple cases, the customary way of obtaining approximate solutions to the quantum mechanical many-body problem proceeds through the minimization of the variational functional

$$E_v^N[\Psi^N] = <\Psi^N|\widehat{H}_v^N|\Psi^N> \quad \text{with} \quad <\Psi^N|\Psi^N> = 1 \quad (2)$$

by carrying out variations of $\Psi^N \in \mathcal{L}_N$ (where $\mathcal{L}_N$ is the N-particle antisymmetric Hilbert space):

$$E_0^{v,N} = \inf_{\Psi^N \in \mathcal{L}_N} \left\{ E_v^N[\Psi^N] \right\}. \quad (3)$$

Since the variational inequality $<\Psi^N|\widehat{H}_v^N|\Psi^N> \geq E_0^{v,N}$ holds for $\Psi^N \in \mathcal{L}_N$ the values of $E_v^N[\Psi^N]$ belong to the set

$$\mathcal{E}^{v,N} = \{E_v^N : E_v^N \geq E_0^{v,N}\}. \quad (4)$$

A Hamiltonian containing up to two-particle operators, such as the one of Eq. (1) can be written as

$$\widehat{H}_v^N = \sum_{i=1}^{N-1}\sum_{j=i+1}^{N} \widehat{K}_v^N(\mathbf{r}_i, \mathbf{r}_j), \quad (5)$$

where the reduced two-particle Hamiltonian is $\widehat{K}_v^N(\mathbf{r}_i, \mathbf{r}_j) = \widehat{K}_0^N(\mathbf{r}_i, \mathbf{r}_j) + \widehat{K}_{ext}^N(\mathbf{r}_i, \mathbf{r}_j)$ and where

$$\widehat{K}_0^N(\mathbf{r}_i, \mathbf{r}_j) = \frac{1}{(N-1)}\left[\frac{-1}{2}\left(\nabla_{\mathbf{r}_i}^2 + \nabla_{\mathbf{r}_j}^2\right)\right] + \frac{1}{|\mathbf{r}_i - \mathbf{r}_j|}, \quad (6)$$

and

$$\widehat{K}_{ext}^N(\mathbf{r}_i, \mathbf{r}_j) = \frac{1}{(N-1)}\Big(v(\mathbf{r}_i) + v(\mathbf{r}_j)\Big). \quad (7)$$

Defining the pure-state N-particle density matrix operator as

$$\hat{D}^N_{\Psi^N}(\mathbf{r}'_1, ..., \mathbf{r}'_N; \mathbf{r}_1, ..., \mathbf{r}_N) = |\Psi^{N*}(\mathbf{r}'_1, ..., \mathbf{r}'_N) >< \Psi^N(\mathbf{r}_1, ..., \mathbf{r}_N)| \quad (8)$$

then, the reduced 2-matrix operator is

$$\begin{aligned}\hat{D}^2_{\Psi^N}(\mathbf{r}_1, \mathbf{r}_2; \mathbf{r}'_1, \mathbf{r}'_2) &= \frac{N(N-1)}{2} \int d\mathbf{r}_3.. \int d\mathbf{r}_N \hat{D}^N_{\Psi^N}(\mathbf{r}'_1, .., \mathbf{r}'_N; \mathbf{r}_1, .., \mathbf{r}_N) \\ &\equiv \hat{L}^N_2 \hat{D}^N_{\Psi^N}\end{aligned} \quad (9)$$

where we denote the contraction operator by $\hat{L}^N_2$. ¿From here it follows that

$$E^N_v[\Psi^N] = Tr_2[\hat{K}^N_v \hat{D}^2_{\Psi^N}], \quad (10)$$

where

$$Tr_2[\hat{K}^N_v D^2_{\Psi^N}] = \int d\mathbf{r}_1 \int d\mathbf{r}_2 \hat{K}^N_v(\mathbf{r}'_1, \mathbf{r}'_2) D^2_{\Psi^N}(\mathbf{r}_1, \mathbf{r}_2; \mathbf{r}'_1, \mathbf{r}'_2)\Big|_{\mathbf{r}'_1 = \mathbf{r}_1, \mathbf{r}'_2 = \mathbf{r}_2}. \quad (11)$$

Equation (10) states the equivalence between the energy functionals $E^N_v[\Psi^N]$ and $Tr_2[\hat{K}^N_v \hat{D}^2_{\Psi^N}]$.

It then follows from Eq. (10), that the variational principles given by Eqs. (3) can be rewritten as

$$\begin{aligned}E^{v,N}_0 &= \inf \left\{ Tr_2[\hat{K}^N_v \hat{D}^2_{\Psi^N}] \right\} \\ \hat{D}^2_{\Psi^N} &= \hat{L}^N_2 \hat{D}^N_{\Psi^N}. \\ \hat{D}^N_{\Psi^N} &\in \mathcal{P}_N\end{aligned} \quad (12)$$

In this equation, $\mathcal{P}_N$ is the set of pure-state N-matrix operators:

$$\mathcal{P}_N = \{\hat{D}^N = |\Psi^N >< \Psi^N| : Tr_N \hat{D}^N = 1; (\hat{D}^N)^2 = \hat{D}^N; \Psi^N \in \mathcal{L}_N\}. \quad (13)$$

For completeness, and since we will use it below, we also define the set $\mathcal{E}_N$ of ensemble N-matrix operators:

$$\mathcal{E}_N = \{\hat{D}^N = \sum_i w_i \hat{D}^N_i : \hat{D}^N_i = |\Psi^N_i >< \Psi^N_i|; 0 \leq w_i \leq 1;$$

$$Tr_N \hat{D}^N = 1; \Psi^N_i \in \mathcal{L}_N\}. \quad (14)$$

It is also possible to rewrite the variational principle in terms of the 1-matrix $\hat{D}^1_{\Psi^N}$. This can be accomplished by introducing the reduced one-particle operator $\hat{h}([\hat{D}^2_{\Psi^N}]; \mathbf{r}_1, \mathbf{r}'_1)$ defined as follows:

$$\hat{h}^N_v([\hat{D}^2_{\Psi^N}]; \mathbf{r}_1, \mathbf{r}'_1)\hat{D}^1_{\Psi^N}(\mathbf{r}_1, \mathbf{r}'_1) \equiv \int d2 \hat{K}^N_v(\mathbf{r}'_1, \mathbf{r}'_2)\hat{D}^2_{\Psi^N}(\mathbf{r}_1, \mathbf{r}_2; \mathbf{r}'_1, \mathbf{r}'_2)\Big|_{\mathbf{r}'_2 = \mathbf{r}_2}. \quad (15)$$

The variational principle then becomes

$$E_0^{v,N} = \inf \left\{ Tr_1\left[\hat{h}([\hat{D}^2_{\Psi^N}];\mathbf{r}_1,\mathbf{r}'_1)\hat{D}^1_{\Psi^N}(\mathbf{r}_1,\mathbf{r}'_1)|_{\mathbf{r}_1=\mathbf{r}'_1}\right]\right\}. \quad (16)$$

$$\hat{D}^2_{\Psi^N} = \hat{L}^N_2 \hat{D}^N_{\Psi^N}$$

$$\hat{D}^1_{\Psi^N} = \hat{L}^N_1 \hat{D}^N_{\Psi^N}$$

$$\hat{D}^N_{\Psi^N} \in \mathcal{P}_N$$

In Eq. (16), in addition to the conditions that the 1- and -2 matrices come from an N-matrix defined in the set $\mathcal{P}_N$, we see that still another condition must be satisfied, namely, that the reduced one-particle operator $\hat{h}^N_v(\mathbf{r}_1,\mathbf{r}'_1)$ be an implicit function of the N-representable 2-matrix. Although it would seem difficult to meet this condition, the reason why it is desirable to recast the many-body problem in terms of the 1-matrix is that the necessary and sufficient conditions of N-representability of the ensemble 1-matrix are known [1]. It has been shown that these ensemble N-representability conditions can be used in order to solve the problem for the pure-state 1-matrix. This is done, for example, in the work of Nguyen-Dang et al. [2] and Ludeña [3, 4], where built-in N-representabilty conditions are introduced in the variational principle so as to guarantee a proper construction of the one-particle operator $\hat{h}([\hat{D}^2];\mathbf{r}_1,\mathbf{r}'_1)$.

In Section 2, we deal discuss the problem of functional N-representability for the energy functionals given in terms of the 2- and 1-matrices [5]. In Section 3, we illustrate the application of functional N-representability criteria for the construction of the reduced single particle operator $\hat{h}^N_v(\mathbf{r}_1,\mathbf{r}'_1)$. In Section 4, we present results of an exact N-representable functional for Hooke's atom. In Section 5 we present some conclusions.

2. Functional N-representability

The basic aspect of the reduced matrix formulation of quantum mechanics is that it bypasses wavefunctions and relies entirely on reduced matrices [1, 6, 7]. Thus, although the variational principle for the 2-matrix given by Eq. (12) and for the 1-matrix by Eq, (16), are in principle correct, they do not fulfill the aim of reduced matrix theory as the conditions are expressed in terms of wavefunctions and of the 1- and 2-matrices obtained from wavefunctions.

For the case of 2-matrices the problem can be reformulated by introducing the set $\mathcal{P}^2_N$ of all N-representable 2-matrices. Clearly, if no reference is made to wavefunctions, the set $\mathcal{P}^2_N$ must be given an **intrinsic** characterization containing all necessary and sufficient conditions for N-representability of the 2-matrix [1, 6, 8, 9, 11, 12, 13, 14, 15, 16, 17].

Unfortunately, no such complete characterization is available at present, as attested by several calculations [18, 19, 20, 21, 22, 10, 23, 24, 25, 26, 27, 28, 29, 30]. Other alternative treatments involving the the 2-matrix have also been advanced [31, 32, 33, 34, 35, 36, 37, 38].

In terms of the set $\mathcal{P}_N^2$, the variational problem (12) is then given by

$$E_0^{v,N} = \inf_{\widehat{D}^2 \in \mathcal{P}_N^2} \left\{ Tr[\widehat{K}_v^N \widehat{D}^2] \right\}. \tag{17}$$

The integrand of $Tr[\widehat{K}_v^N \widehat{D}^2]$ (see Eq. (11)) is the product of the reduced two-particle operator $\widehat{K}_v^N$, which in completely defined for a Hamiltonian of the type given by Eq. (1). Thus, the only condition required on the functional $Tr[\widehat{K}_v^N \widehat{D}^2]$ to belong to the set $\mathcal{E}^{v,N}$ given by Eq. (4) is that $\widehat{D}^2 \in \mathcal{P}_N^2$.

Similarly, let us introduce the set $\mathcal{P}_N^1$ containing, again, all the **intrinsic** necessary and sufficient conditions to render D^1 pure-state N-representable. Once more, except for very simple cases, no such definition is as yet available for this set [39]. On the other hand, the definition of the set $\mathcal{E}_N^1$ stating all the necessary and sufficient conditions that make D^1 ensemble N-representable, is known. In fact, this set is simply defined as:

$$\mathcal{E}_N^1 = \{D^1 : 0 \leq D^1 \leq qI\,;\, Tr_1[D^1] = N\} \tag{18}$$

where q is the highest eigenvalue of the operator D^1 [40].

Let us now consider the variational problem presented in terms of the 1-matrix.

$$E_0^{v,N} = \inf_{\substack{\widehat{D}^2 \in \mathcal{P}_N^2 \\ \widehat{D}^1 \in \mathcal{P}_N^1 \subset \mathcal{E}_N^1}} \left\{ Tr_1\left[\hat{h}([D^2]; \mathbf{r}_1, \mathbf{r}_1') \widehat{D}^1(\mathbf{r}_1, \mathbf{r}_1')|_{\mathbf{r}_1 = \mathbf{r}_1'}\right] \right\}. \tag{19}$$

In this expression, reference to wavefunctions has been dropped and instead the requirement that $D^2 \in \mathcal{P}_N^2$ and that $D^1 \in \mathcal{P}_N^1 \subset \mathcal{E}_N^1$ have been employed.

By functional N-representability we mean the conditions that must be satisfied both by the reduced p-operator and the reduced p-matrix (for $p = 1, 2$) such that the functional belongs to the set $\mathcal{E}^{v,N}$. Thus, in the case of $p = 2$, since the reduced two-particle operator (for the case of a Hamiltonian containing only one- and two-particle operators) has no implicit dependence on the wavefunction (or any reduced density matrix), then functional N-representability just reduces to the N-representability of the 2-matrix. (Clearly, this would not be the case if the Hamiltonian contained three-particle operators). On the other hand, for $p = 1$, we see that the reduced one-particle operator shows an implicit dependence on the 2-matrix. Hence, in order that this reduced one-particle operator be functionally N-representable, it must be obtained either via Eq. (15) or by some equivalent method that guarantees the equivalence required by this equation.

3. Functional N-representability through built-in pure-state N-representability conditions

In what follows, we illustrate the construction of a functionally N-representable reduced single-particle operator based on the occupation numbers and the natural orbitals of the 1-matrix. For this purpose, in the spirit of Levy's constrained search approach [41], let us consider the following ensemble and pure-state functionals of the 1-matrix (where, again, we emphasize their N dependence):

$$F_e^N[D^1] = \inf\{Tr[\widehat{H}_0^N D^N] : D^N \in (\widehat{L}_1^N)^{-1} D^1 \cap \mathcal{E}_N ; D^1 \in \mathcal{E}_N^1\} \quad (20)$$

and

$$F_p^N[D^1] = \inf\{Tr[\widehat{H}_0^N D^N] : D^N \in (\widehat{L}_1^N)^{-1} D^1 \cap \mathcal{P}_N ; D^1 \in \mathcal{P}_N^1\} \quad (21)$$

However, since the ensemble and pure-state functionals of the 1-matrix are equivalent over the set of pure state 1-matrices [42, 43, 2],

$$F_e^N[D^1] = F_p^N[D^1] \equiv F^N[D^1] ; D^1 \in \mathcal{P}_N^1 \quad (22)$$

one need only impose the known ensemble N-representability conditions on the 1-matrix, provided that $F^N[D^1]$ is constructed in a particular way so as to guarantee functional N-representability. This can be accomplished, for example, by varying $F^N[D^1]$ subject to the following conditions [2, 3, 4]:

$$\begin{align}
(1) \quad & Tr[D^N] = 1 & (23)\\
(2) \quad & D^N \in (\widehat{L}_1^N)^{-1} D^1 \cap \mathcal{E}_N & (24)\\
(3) \quad & (D^N)^2 = D^N & (25)
\end{align}$$

Note that the fixed $D^1 = \gamma$ satisfies the ensemble conditions given by Eq. (18). Conditions (1) and (2) lead to the the auxiliary functional [2]:

$$G[D^N] = Tr[\widehat{H}_0^N D^N] - E\left(tr[D^N] - 1\right) - Tr_1\left[\alpha\left(L_1^N D^N - \gamma\right)\right] \quad (26)$$

where E is the Lagrange multiplier incorporating condition (1) and $\alpha(\mathbf{r}_1, \mathbf{r}_1')$ is the Lagrange multiplier function accounting for condition (2). Condition (3) basically states that D^N corresponds to a pure-state and leads to the following condition on the allowed variations:

$$\delta D^N = D^N \left(\delta D^N \right) + \left(\delta D^N \right) D^N \tag{27}$$

As a result, one obtains [2]:

$$F^N[\gamma] = Tr[\widehat{H}_0^N D_\gamma^N] = Tr_1\left[\alpha' \gamma \right] \tag{28}$$

where $\alpha'(\mathbf{r}_1, \mathbf{r}_1') = \alpha(\mathbf{r}_1, \mathbf{r}_1') - EI(\mathbf{r}_1)I(\mathbf{r}_1')$ and D_γ^N is the N-matrix that yields the fixed γ. In particular, when the fixed $D^1 = \gamma$ is given in terms of the natural orbital expansion

$$\gamma(\mathbf{r}_1, \mathbf{r}_1') = \sum_{i=1}^{m} \lambda_i \chi_i^*(\mathbf{r}_1) \chi_i(\mathbf{r}_1') \tag{29}$$

and the N-matrix is expanded as

$$D^N = \sum_{I,J} C_I^* C_J D_I D_J \tag{30}$$

in terms of the M Slater determinants D_I that can be constructed from a set on m natural orbitals:

$$D_I = (N!)^{-1/2} \det[\chi_{i_1}(\mathbf{r}_1), \chi_{i_2}(\mathbf{r}_2), ..., \chi_{i_N}(\mathbf{r}_N)]; \quad I = \{i_1, i_2, ..., i_N\}, \tag{31}$$

then, the natural orbital occupation numbers satisfy the following $m(m+1)/2$ relations:

$$\lambda_i = \sum_{I \ni i} |C_I|^2 \tag{32}$$

$$0 = \sum_{I \ni i} \sum_{J \ni j} C_I C_J^* \Delta_{IJ}^{ij} \tag{33}$$

where

$$\Delta_{IJ}^{ij} = \begin{cases} (-1)^{k-l} & \text{for } \{I - i\} = \{J - j\}; i = i_k, j = j_l \\ 0 & \text{otherwise} \end{cases} \tag{34}$$

The auxiliary functional becomes [2]:

$$G[\{C_I\}] = \sum_{I,J} C_I C_J^* <D_I|\widehat{H}_0^N|D_J> -E\left(\sum_I |C_I|^2 - 1 \right)$$
$$- \sum_i \alpha_{ii} \left[\sum_{I \ni i} |C_I|^2 - \lambda_i \right] - \sum_{i \neq j} \alpha_{ij} \left[\sum_{I \ni i} \sum_{J \ni j} C_I C_J^* \Delta_{IJ}^{ij} \right] \tag{35}$$

¿From Eq. (35) we obtain upon differentiation with respect to the coefficients C_I^*, the following set of linear equations;

$$C_I\left(<D_I|\hat{H}_0^N|D_I> - \sum_{i\in I}\alpha'_{ii}\right) - \sum_{J\neq I}C_J\left(<D_I|\hat{H}_0^N|D_J>\right.$$
$$\left. - \sum_{i\in I\neq j\in J}\alpha'_{ij}\Delta_{IJ}^{ij}\right) = 0 \qquad (36)$$

This set of M linear equations provides relations between the coefficients and the as yet undetermined matrix $\underline{\alpha}'$. Hence, we obtain the coefficients as the following functions $C_I(\underline{\alpha}', \{\chi_i\})$. Introducing these coefficients into Eqs. (32) and (33), we obtain a set of $m(m+1)/2$ equations for the symmetric matrix $\underline{\alpha}'$. Solving these equations we finally obtain $\underline{\alpha}'$ in terms of the occupation numbers $\{\lambda_i\}$ and the natural orbitals $\{\chi_i\}$. Finally, in view of Eq. (28), we can obtain the functionally N-representable expression for the internal energy functional $F^N[D^1 = \gamma]$.

The above results, obtained in 1985 by Nguyen-Dang et al. [2] have particular relevance with respect to the treatment of the 1-matrix problem given by Donnelly and Parr in 1978 [44] and Gilbert in 1975 [45]. For a more recent approach based on natural spin orbitals, see Goedecker and Umrigar [46].

4. Functional N-representability in DFT: Application to Hooke's atom

4.1 General considerations

Density functional theory [47, 48, 49, 50] exploits the fact that the external potential v and the density ρ, as shown by the Hohenberg-Kohn theorem [2], are in a one-to-one correspondence. This implies that through Levy's constrained search [41] a one-to-one correspondence between a functional $F^N[\rho]$ and a given fixed ρ can be established:

$$F^N[\rho] = \inf_{\substack{D_\rho^N \longrightarrow \rho; D_\rho^N \in \mathcal{P}_N \\ \rho \in \mathcal{J}_N}} \left\{Tr[\hat{H}_0^N D_\rho^N]\right\}. \qquad (37)$$

where $\mathcal{J}_N$ is the set of admissible one-particle densities [40] (here again, we emphasize the N-dependence of $F^N[\rho]$, the so-called "universal functional"). As a consequence, the energy can be expressed as:

$$E_0^{v,N} = \inf_{\rho \in \mathcal{J}_N} \left\{E_v^N[\rho]\right\}, \qquad (38)$$

where the total energy expressed as a functional of the one-particle density is

$$E_v^N[\rho] = F^N[\rho] + \int d\mathbf{r}\rho(\mathbf{r})v(\mathbf{r}). \tag{39}$$

The critical aspect of density functional theory has to do with the generation of the functional $F^N[\rho]$. The Hohenberg-Kohn theorem only guarantees the existence of the functional but does not provide a systematic procedure for its construction. On the other hand, application of Levy's constrained search, requires that a workable procedure be developed for scanning over all N-representable N-matrices, $D_\rho^N \in \mathcal{P}_N$ (or equivalently, over all $\Psi_\rho^N \in \mathcal{L}_N$) that yield the fixed density ρ.

The functional $F^N[\rho]$ has been labeled as "universal" in view of the fact that the external potential is absent from its definition. An argument against the full universality of this functional follows from its dependence on N, the number of electrons. According to Lieb [40], "This fact is unavoidable and frequently overlooked".

The condition of functional N-representability on the functional $E_v^N[\rho]$ can be stated as all the necessary and sufficient conditions on $F^N[\rho]$ such that $E_v^N[\rho] \in \mathcal{E}^{v,N}$ (see Eq. (4)).

4.2 Constructive approach based on local-scaling transformations

The same kind of reasoning that we have used in the case of the 1-matrix discussed in the previous Section, may also be employed in the present case and obtain an expression equivalent to Eq. (28) but where all terms are given as explicit functions of the one-particle density. Although, at first sight such an endeavour seems to be unrealizable, in practice, it has been shown to be possible through the application of local-scaling transformations.

Based on these transformations, it is possible to formulate an alternative approach to density functional theory, which, by construction satisfies the functional N-representability condition [52, 53, 54, 55, 56]. Briefly, through the application of local-scaling transformations to the coordinates of a generating wavefunction $\Psi^N \in \mathcal{L}_N$ (whose form can be chosen to satisfy certain physical criteria but which does not have to be calculated) one can generate a density-dependent wavefunction $\Psi_\rho^N \in \mathcal{L}_N$:

$$\Psi_\rho^N(\mathbf{r}_1,...,\mathbf{r}_N) = \prod_{i=1}^N \sqrt{\frac{\rho(\mathbf{r}_i)}{\rho_g(\lambda([\rho],\mathbf{r}_i)\mathbf{r}_i)}} \Psi_g\Big(\lambda([\rho],\mathbf{r}_1)\mathbf{r}_1,...,\lambda([\rho],\mathbf{r}_N)\mathbf{r}_N\Big). \tag{40}$$

where the density-dependent scaling function $\lambda([\rho], \vec{r})$ is defined through the first-order differential equation

$$\lambda([\rho], \vec{r}) = \left(\frac{\rho(\vec{r})}{\rho_g(\lambda([\rho], \vec{r})\vec{r})}\right)^{1/3} \left((1 + \vec{r} \cdot \nabla_{\vec{r}} \ln \lambda([\rho], \vec{r})\right)^{-1/3}. \quad (41)$$

The variational principle becomes

$$E_0^{v,N} = \inf_{\substack{\Psi_\rho^N \in \mathcal{L}_N \\ \rho \in \mathcal{J}_N}} \{E_v^N[\Psi_\rho^N]\} \quad (42)$$
$$\quad (43)$$

where $E_v^N[\Psi_\rho^N] = <\Psi_\rho^N|\widehat{H}_v^N|\Psi_\rho^N>$. Note that this energy functional depends on the locally-scaled wavefunction Ψ_ρ^N and not only on the density ρ. This is a crucial difference with ordinary Kohn-Sham theory, where it is assumed that $E_v^N \equiv E_v^N[\rho]$. In fact, in the local-scaling transformation version of DFT, the energy is a functional of the density and of the generating wavefunction Ψ^N:

$$E_v^N[\Psi_\rho^N] = E_v^N[\rho, \Psi^N]. \quad (44)$$

When the local-scaling transformations are applied to the exact ground-state wavefunction, then we generate the following functional

$$E_v^N[\Psi_\rho^N] = E_v^N[\rho, \Psi_0^{v,N}] \quad (45)$$

which not only depends upon ρ and N but also depends implicitly on the external potential v through the exact wavefunction $\Psi_0^{v,N}$.

4.3 Construction of exact functional for Hooke's atom

The constructive procedure sketched above has been used here in order to generate an explicit and exact functional for Hooke's atom [57, 58, 59, 60, 61, 62, 63, 64]. The point is that because $\lambda([\rho], \vec{r})$ is density-dependent we can rewrite the energy functional $E_v^N[\Psi_\rho^N]$ as an explicit functional of the one-particle density. Since for Hooke's atom, the exact wavefunction is known, we can generate the energy functional by applying local-scaling transformations to this wavefunction.

The spatial part of the exact ground-state wave function for Hooke's atom is:

$$\Psi(\mathbf{r}_1, \mathbf{r}_2) = \frac{1}{2\pi\sqrt{5\pi + 8\sqrt{\pi}}} e^{(-r_1^2/4 - r_2^2/4)} \left(1 + \frac{1}{2}r_{12}\right) \quad (46)$$

Its associated one-electron density is:

$$\rho_\Psi(r_1) = \frac{2}{\left(\sqrt{5\pi + 8\sqrt{\pi}}\right)^2} e^{-r_1^2/2}$$

$$\times \frac{1}{2}\left(8e^{-r_1^2/2} + 7\sqrt{2\pi} + \frac{4\sqrt{2\pi}\mathrm{Erf}(r_1/\sqrt{2})}{r_1}\right.$$

$$\left. + 4\sqrt{2\pi}\mathrm{Erf}(r_1/\sqrt{2})r_1 + \sqrt{2\pi}r_1^2\right) \quad (47)$$

Following the customary procedure employed in the Kohn-Sham version of in DFT, we decompose the total energy as follows:

$$E_v^N[\Psi_\rho^N] = T_s[\{\phi_i^{KS}\},\rho] + E_{Coul}[\rho] + E_{ext}[\rho] + E_x[\{\phi_i^{KS}\},\rho] + E_c^{KS}[\rho,\Psi_0^{v,N}] \quad (48)$$

where $\{\phi_i^{KS}\}_{i=1}^N$ is the set of Kohn-Sham orbitals. The dependence on the Kohn-Sham orbitals in the non-interacting kinetic energy and exchange and correlation functionals of Eq. (48) is eliminated in the case of Hooke's atom (and, in general, in the case of all two electron systems) due to the fact that $\phi^{KS} = \sqrt{\rho}$. For Hooke's atom, the non-interacting kinetic energy $T_s[\rho]$ is the Weizsäcker term

$$T_s[\rho] = \frac{1}{8}\int d^3\vec{r}\, \frac{(\nabla_{\vec{r}}\rho(\vec{r}))^2}{\rho(\vec{r})} \quad (49)$$

and the exchange energy $E_x[\rho]$ is equal to $-(1/2)E_{Coul}[\rho]$. The other terms appearing in Eq. (48) are the Coulomb interaction:

$$E_{Coul}[\rho] = \frac{1}{2}\int d^3\vec{r}_1 \int d^3\vec{r}_2\, \frac{\rho(\vec{r}_1)\rho(\vec{r}_2)}{|\vec{r}_1 - \vec{r}_2|}, \quad (50)$$

the energy due to the external potential:

$$E_{ext}[\rho] = \int d^3\vec{r}\, \rho(\vec{r})\, v_{ext}(\vec{r}), \quad (51)$$

and the Kohn-Sham correlation energy which depends both on ρ and on the exact wavefunction $\Psi_0^{v,N}$. This complicated dependence manifests itself through the fact that the correlation functional involves in its definition the total kinetic energy term $T[\Psi_\rho^N]$ and the electron-electron energy $E_{ee}[\Psi_\rho^N]$. Bearing in mind Eq. (45) it becomes:

$$E_c^{KS}[\rho,\Psi_0^{v,N}] = T[\rho,\Psi_0^{v,N}] - T_W[\rho] + E_{ee}[\rho,\Psi_0^{v,N}] - (1/2)E_{Coul}[\rho] \quad (52)$$

The total kinetic energy can be rewritten as the following density functional:

$$T[\rho, \Psi_0^{v,N}] = T_W([\rho]) + \int_0^\infty dr_1 r_1^2 \Big(\rho^{1/3}(r_1)\rho'(r_1)F_{1/3}([\rho];r_1) \\ + \rho^{5/3}(r_1)F_{5/3}([\rho];r_1)\Big) \quad (53)$$

Similarly, the electron-electron energy functional takes the following form:

$$E_{ee}[\rho, \Psi_0^{v,N}] = \sum_{l=0}^\infty \int_0^\infty dr_1 r_1^2 \rho^{(l+4)/3}(r_1) F_{ee}^{(l)}([\rho];r_1) \quad (54)$$

Both, in Eqs. (53) and (54) we observe that the functionals result from the products of the "universal" terms $\rho^{1/3}(r_1)\rho'$, $\rho^{5/3}(r_1)$, and $\rho(r_1)^{(l+4)/3}$ times their corresponding system-dependent modulating factors. Combining these terms we obtain the following functional for the correlation energy:

$$\begin{aligned}E_c[\rho, \Psi_0^{v,N}] &= \int_0^\infty dr_1 r_1^2 \Big\{\rho^{1/3}(r_1)\rho'(r_1)F_{1/3}([\rho];r_1) \\ &+ \rho^{5/3}(r_1)\Big(F_{5/3}([\rho];r_1) + F_{ee}^{(1)}([\rho];r_1)\Big) \\ &+ \rho^{4/3}(r_1)F_{ee}^{(0)}([\rho];r_1)\Big\} - \frac{1}{2}E_{Coul}[\rho] \\ &+ \sum_{l=2}^\infty \int_0^\infty dr_1 r_1^2 \rho(r_1)^{(l+4)/3} F_{ee}^{(l)}([\rho];r_1) \quad (55)\end{aligned}$$

A more complete discussion of these functionals including explicit expressions for the enhancement factors is given elsewhere [65]

5. Conclusions

We have discussed the problem of functional N-representability of the 1-matrix and of the reduced one-particle operator associated with the 1-matrix in order to illustrate some of the difficulties that arise when one tries to construct a 1-matrix theory that complies with the variational principle. To this effect, we have introduced built-in N-representability conditions that allow for the construction of a functionally N-representable reduced one-particle operator. This complements a previous work where examples were given of non-N-representable and N-representable functionals of the 1-matrix for Hooke's atom [5, 64].

In the present work, we extend this discussion to functionals that depend on the one-particle density and present an exact functional for

Hooke's atom. This functional is strictly variational, viz., for any admissible density it gives an upper bound to the exact Hooke's atom energy. Only when the functional is evaluated with the exact Hooke's atom density, does it yield the exact energy. Moreover, in no case does it go below the exact energy.

In our opinion, the notion that the only N-representability condition that must be satisfied by an energy functional expressed in terms of the one-particle density is that ρ must be N-representable (a condition, which, as was shown by Gilbert [45], can be trivially satisfied), is misleading. The claim we set forth in this article is that in addition, a condition concerning the functional N-representability of $E_v^N[\rho]$ must also be taken into account. Functional N-representability is defined in terms of conditions that must be satisfied by $E_v^N[\rho]$ so that it complies with the variational principle.

The results obtained for the exact Hooke's atom functional are rather enlightening in the sense that one **derives** some of the functional forms usually employed in the expressions of approximate density functionals. Thus, for example, we observe that the exact kinetic energy functional is made up of the Weizsäcker term plus another term which depends on $\rho^{5/3}(r_1)$ times an enhancement factor. This agrees with the usual approximations. But in addition, we find another term whose leading factor is $\rho^{1/3}(r_1)\rho'$. This term is not usually included in approximate functionals but does appear in the present case. This opens the question of whether one should include this type of term in approximate kinetic energy functionals.

In addition, one observes that in the correlation energy functional, there appear terms of the type $\rho(r_1)^{(l+4)/3}$ for $l = 0,...,\infty$. It is particularly interesting in this respect the fact that the exact correlation functional for Hooke's atom contains a term $\rho^{4/3}(r_1)$, which is exactly the Dirac exchange term. This results might contribute to shed light on the fact that GGA exchange functionals seem to be including the non-dynamical correlation effect.

Finally, let us also mention, that the possibility afforded by the present application of local-scaling transformations to construct exact functionals for specific systems, might seem to lead to a contradiction with respect to the existence of a "universal" functional. We hope that the present results will contribute to kindle a discussion concerning this fundamental question in density functional theory.

6. Acknowledgments

We gratefully acknowledge support of FONACIT of Venezuela through Group Project No. G-97000741.

References

[1] A.J. Coleman, Rev. Mod. Phys. **35**, 668 (1963)

[2] T.T. Nguyen-Dang, E.V. Ludeña and Y. Tal, J. Mol. Struct (Theochem) **120**, 247 (1985).

[3] E.V. Ludeña, J. Mol. Struct (Theochem) **123**, 371 (1985).

[4] E.V. Ludeña, in *Density Matrices and Density Functionals*, edited by R.M. Erdahl and V. Smith, Jr. (Reidel, Dordrecht, 1987); p. 289.

[5] E.V. Ludeña, V. Karasiev, A. Artemiev, and D. Gómez, in *Many-Electron Densities and Reduced Density Matrices*, edited by J. Cioslowski (Kluwer Academic/Plenum Publishers, New York, 2000); p. 209.

[6] C. Garrod and J. Percus, J. Math. Phys. (N.Y.) **5**, 1756 (1964)

[7] For a revision, see Section 4.4 of E.S. Kryachko and E.V. Ludeña, *Energy Density Functional Theory of Many-Electron Systems*, (Kluwer, Dordrecht, 1990).

[8] H. Kummer, J. Math. Phys. (N.Y.) **8**, 2063 (1967)

[9] L.J. Kijewski and J.K. Percus, Phys. Rev. **179**, 45 (1969).

[10] C. Garrod, M.V. Mihailovic and M. Rosina, J. Math. Phys. (N.Y.) **16**, 868 (1975)

[11] A.J. Coleman, J. Math. Phys. **13**, 214 (1972)

[12] A.J. Coleman, Rep. Math. Phys. **4**, 113 (1973)

[13] E.R. Davidson, *Reduced Density Matrices in Quantum Chemistry* (Academic, New York, 1976).

[14] A.J. Coleman, Int. J. Quantum Chem. **11**, 907 (1977).

[15] A.J. Coleman, Int. J. Quantum Chem. **13**, 67 (1978).

[16] (J.E. Harriman, Phys. Rev. A **17**, 1249 (1978).)

[17] (J.E. Harriman, Phys. Rev. A **17**, 1257 (1978).)

[18] L.J. Kijewski and J.K. Percus, Phys. Rev. A **2**, 1659 (1970).

[19] J. Simons and J.E. Harriman, Phys. Rev. A **2**, 1034 (1970)

[20] L.J. Kijewski and J.K. Percus, Int. J. Quantum Chem. **5**, 67 (1971).

[21] J. Simons, J. Chem. Phys. **55**, 1218 (1971)

[22] L.J. Kijewski, Phys. Rev. A **9**, 2263 (1974)

[23] M.V. Mihailovic and M. Rosina, Nucl. Phys. **A237**, 221 (1975).

[24] M. Rosina and C. Garrod, J. Comput. Phys. **18**, 300 (1975).

[25] C. Garrod and M.A. Fusco, Int. J. Quantum Chem. **10**, 495 (1976).

[26] C. Garrod, Int. J. Quantum Chem. **13**, 769 (1978).

[27] M. Rosina, Int. J. Quantum Chem. **13**, 737 (1978).

[28] R.M. Erdahl, Rep. Math. Phys. **15**, 147 (1979).

[29] R.M. Erdahl, C. Garrod, B. Golli and M. Rosina, J. Math. Phys. **20**, 1366 (1979).
[30] G. Csányi, S. Goedecker, and T.A. Arias, Phys. Rev. A **65**, 032510 (2002)
[31] H. Nakatsuji, Phys. Rev. A **14**, 41 (1976).
[32] H. Nakatsuji and K. Yasuda, Phys. Rev. Lett. **76**, 1039 (1996).
[33] K. Yasuda and H. Nakatsuji, Phys. Rev. A **56**, 2648 (1997).
[34] C. Valdemoro, Phys. Rev. A **31**, 2114 (1985).
[35] F. Colmenero, C. Pérez del Valle and C. Valdemoro, Phys. Rev. A **47**, 971 (1993).
[36] F. Colmenero and C. Valdemoro, Phys. Rev. A **47**, 979 (1993).
[37] C. Valdemoro, L.M. Tel and E. Pérez-Romero, Adv. Quantum Chem. **28**, 33 (1997).
[38] D.A. Mazziotti, Phys. Rev. A **57**, 41 (1998)
[39] R.E. Borland and K. Dennis, J. Phys. B **5**, 7 (1972).
[40] E.H. Lieb, in *Density Functional Methods in Physics*, edited by R,M, Dreizler and J. da Providencia (Plenum Press, New York, 1985); p. 31.
[41] M. Levy, Proc. Nat. Acad. Sci. USA **76**, 6062 (1979).
[42] S.M. Valone, J. Chem. Phys. **73**, 1344 (1980).
[43] S.M. Valone and J.F. Capitani, Phys. Rev. A **23**, 2127 (1981).
[44] R.A. Donnelly and R.G Parr, J. Chem. Phys. **69**, 4431 (1978).
[45] T.L. Gilbert, Phys. Rev. B **12**, 2111 (1975).
[46] S. Goedecker and C.J. Umrigar, Phys. Rev. Lett. **81**, 866 (1998).
[47] R. G. Parr and W. Yang, *Density-Functional Theory of Atoms and Molecules* (Oxford University Press, Oxford, 1989).
[48] E.S. Kryachko and E.V. Ludeña, *Energy Density Functional Theory of Many Electron Systems* (Kluwer Academic Publishers, Dordrecht, 1990).
[49] N.H. March, *Electron Density Theory of Atoms and Molecules* (Academic Press, London, 1992).
[50] R.M. Dreizler and E.K.U. Gross, *Density Functional Theory*, Springer-Verlag, Berlin, 1990.
[51] P.C. Hohenberg and W. Kohn, Phys. Rev. **B 136**, 864 (1964); P.C. Hohenberg, W. Kohn and L.J. Sham, Adv. Quantum Chem. **21**, 7 (1990).
[52] E.S. Kryachko and E.V. Ludeña, Phys. Rev. **A 43**, 2179 (1991)
[53] E.V. Ludeña and R. López-Boada, Top. Curr. Chem. **180**, 169 (1996)
[54] R. López-Boada, E.V. Ludeña, R. Pino, V. Karasiev, J. Chem. Phys. **107**, 6722 (1997).
[55] E.V. Ludeña R. López-Boada, V. Karasiev, R. Pino, E. Valderrama, J. Maldonado, R. Colle and J. Hinze, Adv. Quantum Chem. **33**, 49 (1999).
[56] E.V. Ludeña R. V. Karasiev, R. López-Boada, E. Valderrama, J. Maldonado, J. Comput. Chem. **20**, 155 (1999).
[57] N.R. Kestner and O. Sinanoglu, Phys. Rev. **128**, 2687 (1962).
[58] M. Taut, Phys. Rev. A **48**, 3561 (1993).

[59] S. Kais, D.R. Hershbach, N.C. Handy, C.W. Murray, and G.J. Laming, J. Chem. Phys. **99**, 417 (1993).

[60] K. Burke, J.P. Perdew, and M. Levy, in: *Modern Density Functional Theory. A Tool for Chemistry*, edited by J.M. Seminario and P. Politzer (Elsevier, Amsterdam, 1995), p. 29.

[61] A. Samanta and S.K. Ghosh, Phys. Rev. A **43**, 6395 (1991).

[62] P.M. Laufer and J.B. Krieger, Phys. Rev. A **33**, 1480 (1986).

[63] C. Filippi, C.J. Umrigar, and M. Taut, J. Chem. Phys. **100**, 1290 (1994).

[64] A. Artemyev, E.V. Ludeña, and V. Karasiev, J. Mol. Struct. (Theochem) **580**, 47 (2002).

[65] E.V. Ludeña, D. Gómez, V. Karasiev, and P. Nieto, Int. J. Quantum Chem. (submitted).

DENSITY-FUNCTIONAL THEORY FOR THE HUBBARD MODEL: NUMERICAL RESULTS FOR THE LUTTINGER LIQUID AND THE MOTT INSULATOR

K. Capelle
Departamento de Química e Física Molecular
Instituto de Química de São Carlos, Universidade de São Paulo,
Caixa Postal 780, São Carlos, 13560-970 SP, Brazil

N. A. Lima, M. F. Silva, and L. N. Oliveira
Departamento de Física e Informática
Instituto de Física de São Carlos, Universidade de São Paulo,
Caixa Postal 369, 13560-970 São Carlos, SP, Brazil

Abstract   We construct and apply an exchange-correlation functional for the one-dimensional Hubbard model. This functional has built into it the Luttinger-liquid and Mott-insulator correlations, present in the Hubbard model, in the same way in which the usual *ab initio* local-density approximation (LDA) has built into it the Fermi-liquid correlations present in the electron gas. An accurate expression for the exchange-correlation energy of the homogeneous Hubbard model, based on the Bethe Ansatz (BA), is given and the resulting LDA functional is applied to a variety of inhomogeneous Hubbard models. These include finite-size Hubbard chains and rings, various types of impurities in the Hubbard model, spin-density waves, and Mott insulators. For small systems, for which numerically exact diagonalization is feasible, we compare the results obtained from our BA-LDA with the exact ones, finding very satisfactory agreement. In the opposite limit, large and complex systems, the BA-LDA allows to investigate systems and parameter regimes that are inaccessible by traditional methods.

1. The Hubbard model and density-functional theory

The Hubbard model is one of the most venerable models of many-body physics. Originally it was proposed as a simplified description of magnetism in transition metals [1, 2]. This required the model to be formulated on a three-dimensional lattice. Interest in the two-dimensional Hubbard model is more recent, and largely due to Anderson's suggestion that it contains the correct minimal requirements needed for describing the cupper-oxide planes in cuprate high-temperature superconductors [3]. In one dimension the Hubbard model has attracted interest mainly because it presents a fascinating phase diagram, including Luttinger liquid and Mott insulating phases which are at the center of much recent work on strongly correlated systems (see, e.g., Refs. [4, 5, 6, 7, 8] for reviews). Independently of the issue of strong correlations, the one-dimensional Hubbard model (1DHM) has aquired additional significance with the recent experimental confirmation of Luttinger liquid behaviour in quasi one-dimensional systems such as carbon nanotubes [9, 10, 11], and quantum wires [12, 13]; systems that offer great potential for applications in the field of nanotechnology.

In its simplest form the 1DHM reads

$$\hat{H} = -t \sum_{<ij>\sigma} c^\dagger_{i\sigma} c_{j\sigma} + U \sum_i c^\dagger_{i\uparrow} c_{i\uparrow} c^\dagger_{i\downarrow} c_{i\downarrow}, \qquad (1)$$

where t is the hopping matrix element, U the on site interaction, and the sum on i, j is restricted to nearest neighbours. Motivated both by the traditional (strong correlations) and more recent (nanotechnology) interest in the 1DHM, we have recently embarked on a density-functional analysis of this model and some of its extensions [14, 15].

In principle, one could think of several ways in which density-functional theory (DFT) and the Hubbard model can be brought together. One attractive possibility is to use *ab initio* DFT in order to calculate the parameters t and U from first principles. Examples of this approach are Refs. [16, 17]. Alternatively, one can try to incorporate the physics of the Hubbard-model into approximate density functionals of *ab initio* DFT. An example of this line of thought is the so-called LDA+U method, in which a Hubbard U is introduced into the *ab inito* LDA functional [18, 19]. Still another possibility is to use the Hubbard model as a laboratory in which formal questions of DFT (such as the meaning of the Kohn-Sham eigenvalues or the band-gap problem) can be studied. This kind of study was pioneered by Gunnarsson and Schönhammer [20, 21, 22]. Finally, one can consider the Hubbard Hamiltonian (1) as a many-body problem in its own right (and a quite difficult one at that), to which DFT can be applied as a calculational tool [14]. In this paper we are concerned with the latter two possibilities.

A prerequisite for applying DFT to the Hubbard Hamiltonian is, of course, a reformulation of the Hohenberg-Kohn theorem and the Kohn-Sham equations, which were originally formulated for the *ab inito* Hamiltonian and not for model Hamiltonians. This reformulation was accomplished by Gunnarsson and Schönhammer [20, 21], who set up so-called *site-occupation DFT*, in which the occupation number n_i of site i plays the same role as the particle density $n(\mathbf{r})$ does in *ab initio* DFT. Recently we have constructed an explicit and simple LDA-like density functional for the 1DHM, which can be used in conjunction with the Gunnarsson-Schönhammer form of the Kohn-Sham equations [14, 15]. Here, we explain this construction in detail, and present some numerical results for a variety of inhomogeneous Hubbard models.

Both the philosophy and the technical details of the construction of the 1DHM functional are very similar to those of the *ab initio* LDA. To appreciate this similarity, let us briefly recall the construction of the latter. First, one considers a homogeneous interacting electron gas (a charged Fermi liquid) and calculates its total energy as a function of the particle density. This calculation was first performed using perturbation theory [23, 24], and more recently with Quantum-Monte Carlo (QMC) [25]. Next, one subtracts the noninteracting kinetic energy and the Hartree energy to extract the exchange-correlation (xc) energy. The result of these two steps is a numerically defined xc functional. For practical applications one needs a sufficiently simple and simultaneously reasonably accurate parametrization of the numerical data. Such parametrizations are constructed taking into account known exact results, such as high-density and low-density limits or scaling properties [23, 24, 26, 27, 28]. Finally, the resulting analytical expression for the per-volume xc energy of the homogeneous electron gas, $e_{xc}(n)$, is used locally to approximate the xc energy of the inhomogeneous real system,

$$E_{xc}^{LDA}[n] = \int d^3r \, e_{xc}^{hom}(n)|_{n \to n(\mathbf{r})}. \tag{2}$$

In this way one transfers the correlations present in one's reference system, the charged Fermi liquid, into the DFT description of the real inhomogeneous system under study.

Let us now consider the one-dimensional Hubbard model. Here the homogeneous reference system is given by Eq. (1). This model is known [4, 29] to describe a charged Luttinger liquid (except for the case of a half-filled band, where it is a Mott insulator). In comparison with the *ab initio* case, where one needs perturbation theory or QMC to calculate the energy of the reference system, the situation for the 1DHM is rather more favorable, since an *exact* solution is available. This solution is obtained by means of the Bethe Ansatz (BA) [5, 30, 31]. The BA results in a set of coupled integral equations, which must still be solved numerically and simplify only in the $U \to \infty$ and $U \to 0$ limits. In these limits, however, one can extract analytical formulae for the total energy as a function of the site occupation numbers n_i and U. We use these formulae, together with a similar expression valid at any U for precisely half filling (all $n_i = 1$), to construct a simple expression for the per-site total energy of the homogeneous 1DHM. The following steps are then precisely as in the *ab initio* case: we subtract kinetic and Hartree energies and use the result to locally approximate the xc energy of inhomogeneous Hubbard models,[1]

$$E_{xc}^{LDA}[n] = \sum_i e_{xc}^{hom}(n)|_{n \to n_i}. \tag{3}$$

Our expression for $e_{xc}(n)$ is derived in the next section, and numerical results obtained from it are shown in the remainder of this paper. Just as in the *ab initio* case, DFT allows us to study inhomogeneous interacting systems by diagonalizing the Hamiltonian of a noninteracting (Kohn-Sham) system. Interesting inhomogeneities include boundaries, surfaces, charge- and spin-density waves, dimerization, impurities, etc. Most of these can be modeled by adding an additional on-site potential

$$\hat{V} = \sum_{i,\sigma} v_i c_{i\sigma}^\dagger c_{i\sigma} \tag{4}$$

to the homogeneous Hamiltonian (1), others require more complicated extra terms. Some examples are given below.

Before we enter these details, however, we briefly mention two alternative LDA-type functionals that have been applied to the 1DHM. One, below called the 'pseudo-LDA', is a discretized form of the three-dimensional *ab initio* LDA [20, 21]. This functional has not performed very well numerically [15, 32] and has been criticized also on the fundamental grounds that the three-dimensional electron gas, on which the *ab initio* LDA is based, is not the correct reference system for the one-dimensional Hubbard model [15, 33, 34]. The other is a numerical Bethe-Ansatz based LDA [22]. This approach is conceptually similar to ours,

---

[1] Due to the particular form of the on-site interaction only electrons with opposite spins interact in the Hubbard model, so that there is no exchange energy in the quantum-chemical sense. Nevertheless, we use the expressions 'exchange-correlation energy' and 'exchange-correlation functional', in order to emphasize that these are the direct Hubbard counterparts to the corresponding *ab initio* concepts.

but has not resulted in an explicit density functional. Instead, it relies on numerical solution of the Lieb-Wu integral equations for the ground-state energy; and subsequent numerical differentiation in order to obtain the corresponding potential.

2. Exchange-correlation energy of the Hubbard model

From the exact Bethe-Ansatz solution to the homogeneous 1DHM [31] one can extract analytical expressions for the total ground state energy in several important limiting cases. For infinitely strong interactions ($U \to \infty$) and a less than half-filled band ($n < 1$), e.g., one has [5]

$$e(n, t, U \to \infty) = -\frac{2t}{\pi} \sin(\pi n) \qquad (5)$$

where $n = N/L$ is the band filling (a constant in the homogeneous case), N the number of electrons and L the number of lattice sites. In the absence of interactions ($U = 0$) one straightforwardly obtains

$$e(n, t, U = 0) = -\frac{4t}{\pi} \sin\left(\frac{\pi}{2} n\right). \qquad (6)$$

Finally, for an exactly half-filled band ($n = 1$) and any interaction U [5]

$$e(n = 1, t, U) = -4t \int_0^\infty dx \frac{J_0(x) J_1(x)}{x(1 + \exp(xU/2t))}, \qquad (7)$$

where J_0 and J_1 are zero and first order Bessel functions.

We now employ these three results to set up an interpolation formula for intermediate values of U and n. Motivated by the similarity of the limiting expressions (5) and (6) we adopt the functional form

$$e(n, t, U) = -\frac{2t\beta(U/t)}{\pi} \sin\left(\frac{\pi}{\beta(U/t)} n\right), \qquad (8)$$

where β is a function of the ratio U/t. Clearly, for $U \to \infty$ one must have $\beta = 1$, while for $U = 0$ one must recover $\beta = 2$. To fix β for intermediate values of U/t we employ Eq. (7) and determine β from requiring that

$$-\frac{2t\beta(U/t)}{\pi} \sin\left(\frac{\pi}{\beta(U/t)}\right) = -4t \int_0^\infty dx \frac{J_0(x) J_1(x)}{x(1 + \exp(xU/2t))}, \qquad (9)$$

which guarantees the correct result at half filling ($n = 1$). This requirement in fact determines β for all values of U, including the limiting cases $U \to \infty$ and $U = 0$, since in these two limits one can calculate the integral analytically and indeed recovers $\beta = 1$ and $\beta = 2$, respectively.

Note that the integral appearing in Eq. (9) does not lead to computational complications: for any given value of U (i.e., for any fixed Hamiltonian) the right-hand side is just a number (the integral in it is easily calculated numerically and converges rapidly). Having determined this number, Eq. (9) is merely a transcendental equation for β, which can be solved by standard methods and has exactly one solution in the physical interval $(U = 0, U \to \infty)$, that is, in the interval $(\beta = 1, \beta = 2)$. This entire calculation takes place outside the self-consistency cycle of DFT.

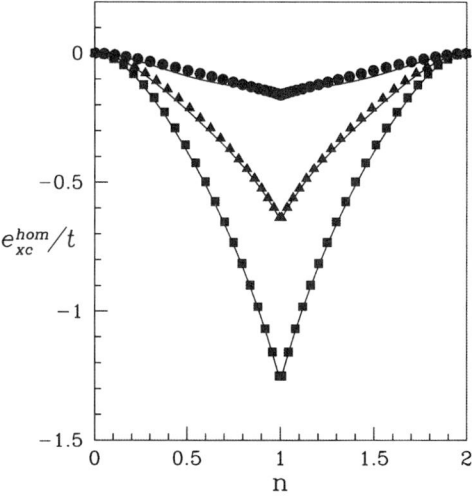

Figure 1. Exchange-correlation energy per site of the homogeneous infinite 1DHM as obtained by numerically solving the Lieb-Wu integral equations resulting from the Bethe Ansatz. Circles: $U = 3$, triangles: $U = 6$, squares: $U = 9$. The full lines are obtained from our expression (8) with (9) and (10). The band filling n ranges from $n = 0$ (empty band) over $n = 1$ (half-filled band) to $n = 2$ (filled band). The form of the curves reflects particle-hole symmetry, and the kinks at $n = 1$ signal the Mott metal-insulator transition.

The interpolation formula (8) with (9) is already our final result for $n \leq 1$, i.e., up to half filling. For a more than half-filled band a particle-hole transformation [31, 5] can be used to express the energy in terms of that for a less than half-filled band,

$$e(n > 1, t, U) = e(2 - n, t, U) + U(n - 1), \tag{10}$$

where $e(2 - n, t, U)$ is the energy for a less than half-filled band, given above. This completes our interpolation of the ground-state energy of the 1DHM. The quality of the expression obtained is illustrated in Fig. 1,

Density-functional theory for the Hubbard model

in which we compare the ground-state energy calculated from Eqs. (8) with (9) and (10) with the one obtained from numerically solving the Lieb-Wu integral equations following from the Bethe Ansatz.

In Fig. 2 we show how our expression interpolates between the $U = 0$ and $U \to \infty$ limits. For comparison purpose we have included in this figure also two curves representing the corrections to the $U \to \infty$ limit up to order $1/U$ and $1/U^3$, respectively [35]. Our interpolation is by construction exact at $U = 0$ and $U \to \infty$. At large U it is very similar to the asymptotic expansions, but unlike these it recovers the correct $U = 0$ limit.

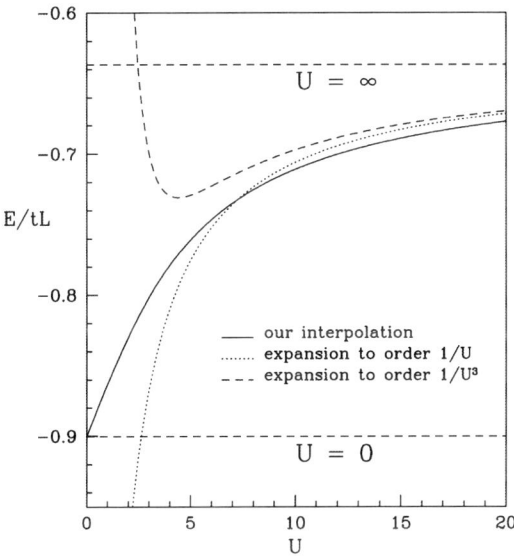

Figure 2. Full curve: total energy of the homogeneous infinite 1DHM with $n = 0.5$, as calculated from our expression (8) with (9) and (10), as a function of interaction strength U. The two dashed horizontal lines denote the limits $U = 0$ and $U \to \infty$, respectively. The dotted and dashed curves are the analytically known [35] correction to the $U \to \infty$ result to order $1/U$ and $1/U^3$, respectively.

Note that here our procedure slightly deviates from the one common in *ab initio* DFT: we have based our expression for $e(n, t, U)$ entirely on the three exact limiting cases, and used the numerical Bethe Ansatz data only for comparison purposes. In the *ab initio* case the LDA functional is based directly on a parametrization of the numerical QMC data, and the exactly known limits serve only as constraints [26, 27, 28]. We could adopt a similar procedure here by introducing a number of free parameters in the functional and fit these to the numerical data. This would clearly provide an even better parametrization of the total energy than the above interpolation, but for our present purposes that interpolation is sufficiently accurate.

Given an expression for the total energy of the homogeneous 1DHM one can use it in either of two ways. First, many interesting observables can be expressed in terms of this energy. If one uses the Bethe Ansatz directly to calculate these, one obtains complicated expressions that must be evaluated numerically and simplify only in the $U = 0$ and $U \to \infty$ limits. On the other hand, a simple approximate expression for $e(n, t, U)$ can be used to obtain simple analytical results also between these limits. An example is the Mott gap that opens at half filling in the homogeneous 1DHM. This gap can be calculated as $\Delta = I - A$, where I and A are ionization energy and electron affinity, respectively. Both of these quantities can be calculated as differences of total ground-state energies. We have recently used our expression for this energy to obtain the following approximate expression for the Mott gap in the thermodynamic limit [15]

$$\Delta(U) = U + 4t \cos\left(\frac{\pi}{\beta(U/t)}\right). \tag{11}$$

This expression can be shown [15] to yield the correct results in the limits $U \to \infty$ and $U = 0$. In between these limits its accuracy is comparable with that of the asymptotic expansion of Δ to order $1/U$ [36]. More details are given in Sec. 3.4 and Ref. [15]. Many other quantities can be treated in the same way. Research along these lines is currently under way in our group.

A second possible use one can make of the expression for $e(n, t, U)$ is to employ it as an input for constructing an LDA-type functional. To this end one must subtract the per-site noninteracting kinetic energy t_s and Hartree energy e_H from $e(n, t, U)$. We define the Hartree energy in general as

$$E_H[n, U] = \frac{U}{4} \sum_i n_i^2, \tag{12}$$

where $n_i = \langle c_{i\uparrow}^\dagger c_{i\uparrow} + c_{i\downarrow}^\dagger c_{i\downarrow} \rangle$ is the local occupation number. (Other definitions are possible, but this one is convenient for our purposes. As long as one uses a consistent expression for the Hartree term in the definition of the xc functional and in the KS equations all choices are equivalent on the exact level.) For a homogeneous system our choice implies $e_H(n, U) = Un^2/4$. The noninteracting kinetic energy is simply given by $e(n, t, U = 0)$, since for $U = v_i = 0$ the Hamiltonian of the 1DHM contains only the kinetic energy term. Our functional then

becomes
$$E_{xc}^{LDA}[n,t,U] = \sum_i e_{xc}(n,t,U)|_{n \to n_i}, \qquad (13)$$

where
$$e_{xc}(n,t,U) = e(n,t,U) - e(n,t,0) - e_H(n,U), \qquad (14)$$

$e(n,t,U)$ is our expression (8) with (9) for $n_i \leq 1$, and use of (10) is implied for $n_i > 1$.

¿From the point of view of formal DFT it is worthwhile to point out that this functional explicitly depends on the interaction U and the hopping parameter t. In fact, this should not come as a surprise: even the *ab initio xc* functional depends on the parameter determining the interaction strength and the coefficients in the kinetic energy operator. The only difference is that in the *ab initio* case these are usually fixed to be e^2 for the interaction and $\hbar^2/2m$ for the kinetic energy, and one does not bother to specify the dependence on e^2 and m in the functionals.

The appearance of these parameters in the density functional is a necessary consequence of the fact that the Hohenberg-Kohn theorem asserts universality of the functional with respect to the external potential $\hat{V}$, but not with respect to the interaction law or the form of the kinetic energy. Even in *ab initio* DFT one requires new functionals when one considers, e.g., phonon-induced electron-electron interactions (such as in DFT for superconductors [37]), or the Dirac kinetic energy (such as in relativistic DFT [38]). In the present, Hubbard, case the dependence of the functional on U and t is strongly constrained: if energies are scaled by t, then all properties of the Hubbard Hamiltonian, including the *xc* functional, depend only on the ratio U/t. Our interpolation, as constructed above, respects this condition. Below we thus follow the universally adopted convention to take t as our unit of energy, and do not include it explicitly among the parameters in the functional.

A final remark on our functional is that it is, of course, far from optimal, and offers much opportunity for improvement. Among other things one could consider to (i) use spin-resolved densities instead of the charge density, i.e., construct an LSDA instead of an LDA, (ii) parametrize the integral (7) to obtain a closed expression for $\beta(U)$, (iii) use scaling conditions to separate exchange and correlation contributions to e_{xc}, (iv) develop a parametrization directly for the numerical data obtained from the Lieb-Wu integral equations, (v) apply self-interaction corrections, (vi) extend the interpolation to negative values of U (interesting in connection with purely electronic superconductivity), etc. Several of these projects are currently under study in our group.

3. Applications

3.1 Luttinger liquids

'Luttinger liquid' is the name usually given to one-dimensional Fermi liquids [4]. One-dimensional metals, for which the Luttinger liquid is the unifying paradigm, behave in many ways so differently from their higher dimensional counterparts (e.g., they do not have low-lying fermionic quasi particles) that a whole new set of concepts and a very specific terminology has been developed to deal with them [4, 5, 29]. Although one-dimensional metals may appear as a theoretical curiosity, there are many systems, such as carbon nanotubes [9, 10, 11], quantum wires [12, 13], edge states in the fractional quantum Hall effect [39, 40], and quasi one-dimensional organic [41, 42] and inorganic [43, 44, 45] conductors, for which experiments indicate Luttinger liquid behaviour. The recent upsurge of interest in nanoscale physics has brought in particular quantum wires and carbon nanotubes into the focus of mainstream research, and as a consequence Luttinger liquid theory has aquired a quite unexpected relevance for device technology and related applications.

The Luttinger model (from which the universality class of one-dimensional metals derives its name) is perhaps the simplest model whose low-energy degrees of freedom are described by Luttinger-liquid phenomenology. For our purposes, however, it is more important that the homogeneous 1DHM for band-fillings $n \neq 1$ (i.e., off half filling) is also a Luttinger liquid. Our functional can be directly applied to this phase of the 1DHM, and used to extract a variety of observables. Here we just consider the ground-state energy; results for other quantities will be published separately.

In Fig. 3 we plot the total energy of a homogeneous 1DHM with $U = 6$ as a function of lattice size, both for open (Hubbard chains) and periodic (Hubbard rings) boundary conditions. For even L we take $N = L/2$ (so that $n = N/L = 1/2$, corresponding to quarter filling). For odd L we take $N = (L-1)/2$. BA-LDA results, obtained from self-consistent solution of the 1DHM Kohn-Sham equations with our functional, are compared with numerically exact ones. The exact (Lanczos) diagonalization was carried up to $L = 14$ on a small PC, taking into account the conservation of particle number and of the z-component of the spin, and advantage of sparse matrix techniques. With a supercomputer, and exploiting symmetries such as total spin, particle-hole symmetry, etc. one can perhaps double the maximum attainable size, but the exponentially increasing Hilbert space makes such calculations prohibitively expensive, and with today's computing technology there is no way of attaining, e.g., $L = 100$.

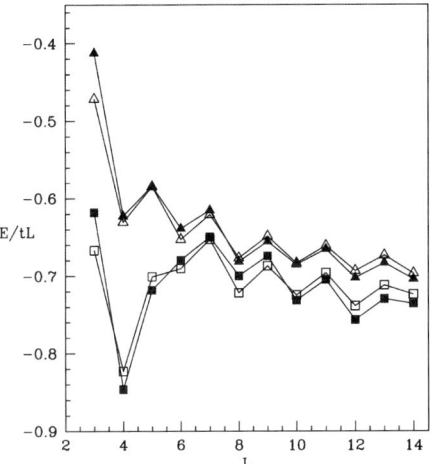

Figure 3. Total energy of a finite 1DHM in the Luttinger liquid phase for $U = 6$. Full triangles: BA-LDA results for open boundary conditions. Open triangles: exact results for open boundary conditions. Full squares: BA-LDA results for periodic boundary conditions. Open squares: exact results for periodic boundary conditions. The lines are only a guide for the eye.

The following conclusions can be drawn from these calculations: (i) Even for very small systems the BA-LDA is reliable. It faithfully reproduces the difference between both types of boundary condition and the even-odd oscillations as a function of the number of lattice sites, with an error of not more than a few percent. In view of the thermodynamic limit built into the reference system for the LDA, this good performance even for small systems is a welcome surprise. (ii) The difference between the exact and the LDA results for moderate L is largely due to the fact that our expression (8) for the energy of the homogeneous reference system is only an interpolation between exact results, but not itself exact for all values of the parameters. A better parametrization (work on which is in progress) would presumably diminish the remaining differences between the exact and LDA curves. (iii) The difference between results obtained for open and for closed boundary conditions only becomes small when the system size exceeds the range accessible with exact diagonalization. The common tenet of solid-state physics 'boundary-conditions do not matter for bulk phenomena' thus is only true when the bulk is already too large for exact calculations to be viable. (iv) The computational effort for the BA-LDA is orders of magnitude lower than for the exact diagonalization, since one must only diagonalize a noninteracting Hamiltonian. In fact it is no problem at all to calculate the energy and other observables for hundreds of sites on a small PC.

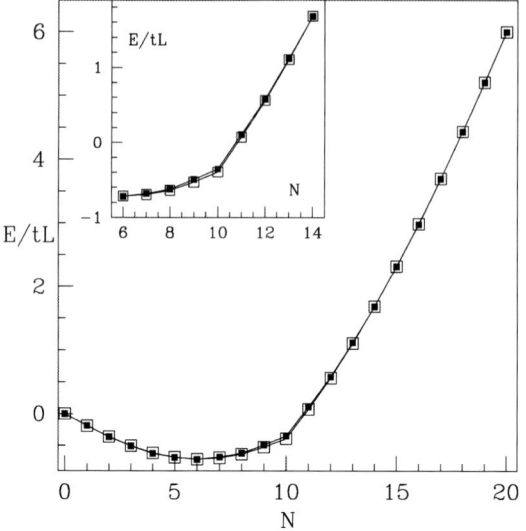

Figure 4. Full squares: total energy of a 1DHM with open boundary conditions, $U = 6$ and $L = 10$, calculated from the BA-LDA as a function of the total number of electrons. Open squares: exact results obtained by numerically diagonalizing the 1DHM. The inset is a zoom into the region near $N = L$, where the BA-LDA functional is discontinuous and the 1DHM undergoes its metal-insulator transition. The lines are a guide for the eye.

Density-matrix renormalization group (DMRG) [46] calculations are capable of attaining similar system sizes, and usually achieve much better accuracy. However, DMRG calculations are encounter difficulties for periodic boundary conditions, are hard to apply to inhomogeneous systems (see below), and become computationally expensive for L in the hundreds or larger. BA-LDA, on the other hand, is less accurate, but does not suffer from either of these drawbacks. A BA-LDA calculation can thus provide useful complementary information to a DMRG one.

As another illustration of the BA-LDA applied to the Luttinger-liquid phase, we plot, in Fig. 4, the energy for a fixed system size $L = 10$ and vary the number of electrons N. Again, exact and BA-LDA results agree well. The abrupt change of slope at $N = L$ is a signal of the metal-insulator transition taking place at half filling.

3.2 Impurity models

In the examples of the previous section the inhomogeneity arose only from the finite size of the system, which was still homogeneous in the bulk. More interesting both from a fundamental and a practical point of view, are systems that are inhomogeneous also in the bulk. An example for such an inhomogeneity that has been much studied in the literature is that of an impurity in a Luttinger liquid. The effect of an impurity on a one-dimensional system is much more profound than on a three dimensional system, because for open boundary conditions the impurity effectively splits the system in two subsystems. Particles can get from one subsystem to the other only by passing through the impurity site. This is to be compared with the situation in a three-dimensional system, in which particles can circumvent the impurity site in many ways. Similarly, for periodic boundary conditions in one dimension there is no closed path that does not involve the impurity, while there are many in higher dimensions.

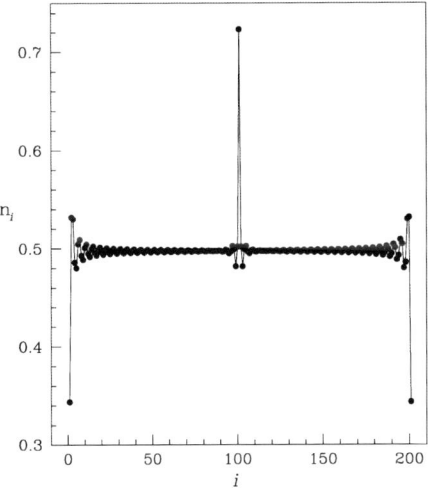

Figure 5. Density distribution for an impurity in a Luttinger liquid with $U = 6$, $n = 0.5$, and open boundary conditions. The impurity is described by an on-site potential of unit strength at the central site. The Friedel oscillations arising at the impurity are clearly visible, and comparable in size with those originating ate the surfaces. The lines are a guide for the eye.

One consequence of this different physics in one dimension is very pronounced Friedel oscillations arising around the impurity. Another is that the convergence to the thermodynamic limit is significantly slowed down by the presence of the impurity. We have discussed these issues in Ref. [14]. Here we provide, for illustration, a plot of the density distribution around the impurity site. Fig. 5 clearly displays the Friedel oscillations arising from the impurity and from the boundary (which in a finite system with open boundary conditions also acts as an impurity).

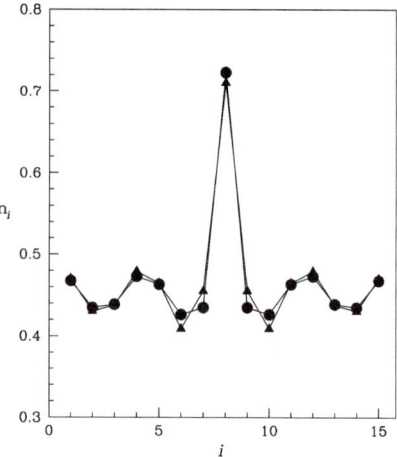

Figure 6. Density distribution of a 15-site 1DHM with an inpurity of strength $v_I = -1$ on the central site ($U = 5$, $n = 1$, periodic boundary conditions). Circles: exact result. Triangles: BA-LDA. The lines are a guide for the eye.

In Fig. 6 we consider a system with $L = 15$ sites, for which exact diagonalization is still possible, and compare the BA-LDA results for the density oscillations with the exact ones. For such small systems the LDA is not expected to do well (recall that it was, just as any other LDA, based on the thermodynamic limit and assumes slow spatial variation of the density). However, in spite of this caveat the LDA density distribution is seen to agree quantitatively with the numerically exact one.

For larger systems, such as that of Fig. 5, exact diagonalization becomes prohibitive. For not too large systems one can compare with DMRG, but even that method is numerically very expensive for impurities in the bulk. As a computationally less expensive alternative one can place the impurities at the boundaries. A systematic DMRG studies of such boundary fields in the 1DHM has been performed in Ref. [47]. For the BA-LDA approach the limits on system size are much less restraining than for other methods, and it makes little difference where one places the impurity. As an example for a calculation with boundary fields, we display here, in Fig. 7, the density at site $i = 1$ for the case in which impurities are located at the boundaries, i.e., at $i = 1$ and $i = L$. In this calculation we have choosen exactly the same parameters as in figure 4 of Ref. [47], so that one can directly compare both results. There are some data points missing for densities very close to $n = 1$. This is due to a convergence problem in the self-consistency cycle, and will be discussed in Sec. 3.4. Apart from this region the BA-LDA curves and the corresponding DMRG curves are in reasonable, but not yet optimal, agreement: the numerical values and global behaviour are very similar, but the local curvature of the BA-LDA results is visibly different from the DMRG one. Improvements on the functional (along the lines described at the end of Sec. 2 and of Sec. 3.4, respectively) are expected to further improve the agreement with the DMRG results.

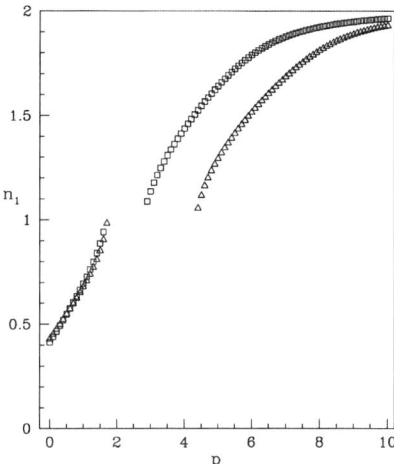

Figure 7. Density at site $i = 1$ for an impurity model in which the impurities of strength p are located at the system's boundary. This plot, obtained with the BA-LDA, is to be compared with Fig. 4 of Ref. [47], obtained using density-matrix renormalization group for the same model. Upper curve (squares): $U = 4$, lower curve (triangles) $U = 6$. In both cases $n = 0.55$ and $L = 100$.

Quite independently of these details, the degree of agreement between the BA-LDA and the exact and DMRG results is remarkable, in view of the fact that the BA-LDA is orders of magnitude faster and less memory-consuming than these more precise methods. The BA-LDA may thus be a useful tool to explore parameter regimes (in particular in large and/or inhomogeneous systems) that are impossible to access with standard methods. An example in which this expanded range of accessible parameter space is immediately useful is the determination of anomalous exponents in Luttinger liquids. These exponents, which govern the asymptotic behaviour of density distributions, correlation functions, etc., are among the parameters that characterize a given Luttinger liquid, and their values and dependence on the interaction can only be extracted from Luttinger liquid theory through some fairly involved mathematics — if at all [4]. Numerical determination of such exponents is hampered by the fact that they become well defined only asymptotically, and systems that are large enough to allow the asymptotic regime to take over are exceedingly hard to treat with traditional methods. The Bethe-Ansatz LDA then provides an attractive alternative. We have given a first example for the determination of such exponents from BA-LDA density distributions in Ref. [14]. A more detailed study is under way.

3.3 Spin-density waves

We now turn to a case in which the inhomogeneity does not occur in real space, but in spin space. Specifically, we add the term

$$\hat{S} = S \sum_{\langle ij \rangle} \left(c_{i\uparrow}^\dagger c_{j\downarrow} + H.c. \right) \tag{15}$$

to the Hamiltonian (1). The *staggered field* S couples spin up and spin down states. For $i = j$ the combination of creation and annihilation operators $c_{i\uparrow}^\dagger c_{i\downarrow} =: \hat{\rho}_{s,i}$ is related to the x and y components of the local spin magnetization via

$$\hat{m}_x(\mathbf{r}) = \mu_0 [\hat{\rho}_{s,i} + \hat{\rho}_{s,i}^\dagger] \tag{16}$$

$$\hat{m}_y(\mathbf{r}) = i\mu_0 [\hat{\rho}_{s,i} - \hat{\rho}_{s,i}^\dagger], \tag{17}$$

where μ_0 is the Bohr magneton. A spin configuration that is not completely specified by the z-component of the full magnetization vector **m**, but requires specification of the x and y-components as well, is noncollinear. Physically, a coupling of the type (15) corresponds to either of the following three cases: (i) A noncollinear ground state, such as the helical or canted spin configurations observed in many rare-earth compounds, the itinerant helical spin-density wave in fcc iron, or domain walls in ferromagnets. (ii) Excitations out of a collinear ground state, such a magnons and solitons in a ferro or antiferromagnet. (iii) A collinear state with the quantization axis chosen to be different from the axis of polarization (e.g., a ferromagnet polarized along the x-axis, but with z chosen as the spin quantization axis). For the present purpose, of developing and testing a DFT for the 1DHM, case (i) is the most interesting one.

In this context the *staggered density*

$$\rho_{s,ij} = \langle \hat{\rho}_{s,ij} \rangle = \langle c_{i\uparrow}^\dagger c_{j\downarrow} \rangle \tag{18}$$

is most conveniently treated by considering it a new fundamental variable, conjugate to the externally applied field S, which enters the formalism on par with the particle density $n_i = \sum_\sigma \langle c_{i\sigma}^\dagger c_{i\sigma} \rangle$. An *ab initio*

DFT based on the corresponding continuous variable

$$\rho_s(\mathbf{r}, \mathbf{r}') = \langle \Psi_\uparrow^\dagger(\mathbf{r}) \Psi_\downarrow(\mathbf{r}') \to, \tag{19}$$

where $\Psi_\sigma(\mathbf{r})$ is a field operator, has been proposed in Ref. [48]. Applications to the Overhauser spin-density wave (SDW) in one and three dimensional electron gases were reported in Ref. [49]. A first application to the 1DHM was presented in Ref. [50], where the stability of the 1DHM to small external staggered fields was investigated.

In this formalism the xc energy becomes a functional of both densities, n and ρ_s. Here we approximate this functional as

$$E_{xc}[n, \rho_s] \approx E_{xc}^{BALDA}[n] - \alpha U \sum_i |\rho_{i,i}|^2, \tag{20}$$

where the first term is the BA-LDA described above, and the second is the $U\alpha$-approximation discussed in Refs. [48, 49, 51]. The coefficient α is an adjustable parameter in the same spirit as in the $X\alpha$ approximation of *ab initio* DFT. Here, however, it does not multiply the usual exchange energy (which is already taken into account by E_{xc}^{BALDA}), but rather the staggered Hartree term that constitutes the driving mechanism for the Overhauser SDW transition in the electron gas. The physical significance of this staggered Hartree term and the coefficient α has been discussed at length in Ref. [49].

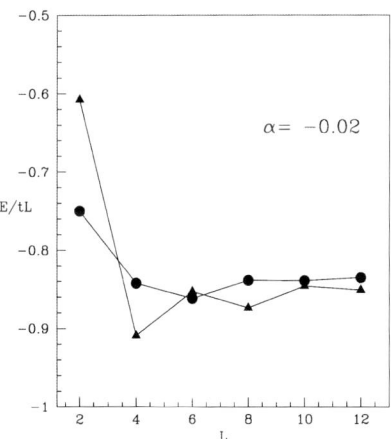

Figure 8. Total energy of a 1DHM with $U = 5$ and $n = 0.5$ in the presence of an external staggered field of strength $S = 0.5$. Circles: numerically exact results. Triangles: BA-LDA+$U\alpha$ approximation with $\alpha = -0.02$.

In Fig. 8 we display the total ground-state energy of a 1DHM with $U = 5$ and an even number of lattice sites, subjected to an external staggered field of strength $S = 0.5$. In the presence of this field the z-component of total spin, S_z is not conserved anymore. One less conservation law makes the exact diagonalization much more demanding in terms of memory usage and computing time, and we present only results for up to $L = 12$ sites. The Kohn-Sham calculations, on the other hand, suffer much less from the lack of conservation of S_z: the appearance of matrix elements connecting up with down states in the Hamiltonian matrix is more than compensated by the absence of particle-particle interactions. In principle these calculations could be carried up to hundreds of sites, but to be able to compare with the exact results we have only gone up to $L = 12$, too.

The best agreement between the exact and the approximate values is achieved for a small negative value of α. The physical significance of this is easily understood in light of the discussion in Refs. [49, 50, 51]: The external staggered field twists the spins away from the z-axis, and the staggered Hartree term always lowers the energy of a noncollinear situation with respect to a collinear one. The factor α corrects the staggered Hartree term, approximately taking into account the correlations not included explicitly in the density functional. A negative value for α means that these correlations overcompensate the energy lowering due to the staggered Hartree term. The staggered density of the system in Fig. 8 is thus exclusively due to the external field S. In other words, the system does not want to accomodate the SDW forced onto it by the external field. This is in agreement with our earlier finding [50] that the unperturbed system (without an external staggered field) does not have an intrinsic instability towards a noncollinear SDW.

Of course, the $U\alpha$ approximation for the ρ_s-dependent part of the functional (20) is much less sophisticated than the BA-LDA for the n-dependent one. However, the quantitative agreement between exact and approximate values, and the physically reasonable sign of the optimal value of α, show that the $U\alpha$ approximation for the ρ_s-dependent part provides at least a useful starting point for further improvements [49]. Concerning the n-dependent part of the functional, we conclude that the BA-LDA remains computationally viable also in situations in which the inhomogeneity arises in spin space, and in which some symmetries are broken.

3.4 Mott insulator

In this section we return to systems that are homogeneous in the bulk, but now we choose the band filling n to be exactly one. For $n = 1$ and $U \neq 0$ the 1DHM is a Mott insulator, i.e., an insulator whose gap arises from correlations, and not as a consequence of the underlying periodic lattice and the resulting single-particle band structure [6, 7, 8]. In DFT one can write the exact many-body gap as a sum of two contributions, $\Delta = \Delta_{KS} + \Delta_{xc}$, where Δ_{KS} is the difference between the highest occupied and the lowest unoccupied single-particle energies, and Δ_{xc} is defined as the discontinuity of the xc potential as a function of the total particle number [52, 53, 54]

$$\Delta_{xc} = \left.\frac{\delta E_{xc}[n]}{\delta n}\right|_{N+\delta} - \left.\frac{\delta E_{xc}[n]}{\delta n}\right|_{N-\delta}, \qquad (21)$$

where $\delta \to 0^+$. The DFT characterization of a pure band insulator is $\Delta = \Delta_{KS}$, while that of a pure Mott insulator is $\Delta = \Delta_{xc}$. In general, of course, both contributions are present simultaneously. Two questions then immediately pose themselves: (1) which of the two is the dominating contribution in a given system, and (2) which of the two is reproduced by common approximate functionals?

As it turns out, the usual LDA and common gradient-corrected functionals do not have any discontinuity, and thus always predict $\Delta_{xc} = 0$. This is the origin of the so-called band-gap problem of DFT. On the other hand, the BA-LDA naturally has a discontinuity at $N = L$, where the underlying homogeneous 1DHM undergoes its metal-insulator transition [31]. Recently we have systematically investigated the resulting Δ_{xc} [15]. In fact, it is simple to calculate Δ_{xc} explicitly in the thermodynamic limit of a homogeneous system, by substituting our expression (14) with (8), (9) and (10) into Eq. (21). The result is precisely our Eq. (11) for the total gap, which was earlier calculated from the difference of ionization energy and electron affinity $I - A$. These latter two quantities can be obtained from ground-state energies according to

$$I = E(N-1) - E(N) \qquad (22)$$
$$A = E(N) - E(N+1). \qquad (23)$$

The agreement between both expressions for Δ does not come as a surprise, because in the thermodynamic limit of a homogeneous 1DHM $\Delta_{KS} \equiv 0$ (the single-particle gap vanishes and a band-structure calculation would predict the system to be a metal), so that $\Delta \equiv \Delta_{xc}$. The fact that we recover this equality shows that our expression for $e_{xc}(n, t, U)$ is reliable enough that both ways of calculating the gap from it lead to the same result.

The importance of the discontinuity in the xc functional is illustrated in Fig. 9, which displays the size of the gap obtained numerically from the BA-LDA [15]; from the continuous pseudo LDA of Refs. [20, 21] (GS-LDA); directly from Eq. (11), which also follows from the BA-LDA, but

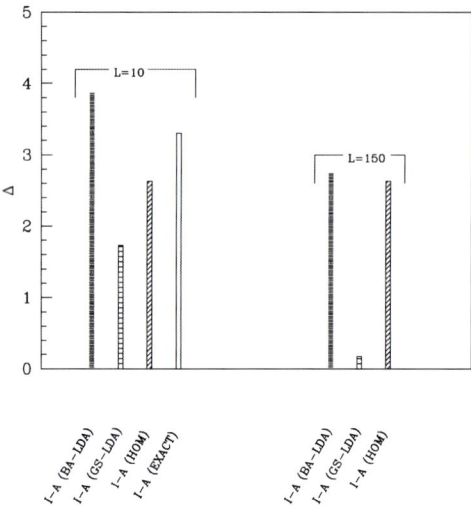

Figure 9. Energy gap of the 1DHM with $U = 6$, calculated for two different lattice sizes, with the methods indicated on the bars.

is valid only for $L \to \infty$; and, for $L = 10$, also from exact diagonalization. Two conclusions can be drawn immediately from these data: (i) The exact gap is well approximated by the BA-LDA gap, but widely underestimated by the pseudo-LDA gap. This is due to the fact that the BA-LDA gap contains an xc contribution due to its discontinuity, in addition to the single-particle gap. (ii) For $L = 150$ sites the asymptotic formula (11) and the numerically determined BA-LDA gap agree already quite well, whereas the pseudo-LDA gap goes to zero as the system size is increased. This latter behaviour is easy to understand: for a functional without a discontinuity, the gap is entirely due to the difference between Kohn-Sham eigenvalues. This difference, i.e., the single-particle gap, is zero in the infinite system, and the Mott metal-insulator transition is exclusively driven by the discontinuity. A more detailed DFT analysis of the Mott gap in the 1DHM can be found in Ref. [15] and, using a somewhat different approach, Ref. [22].

The intrinsic discontinuity of the BA-LDA is of course a very desirable feature for the calculation of energy gaps. The *ab initio* LDA is based on a charged Fermi liquid (a perfect metal), and the local approximation of the xc energy of the inhomogeneous system by that of this homogeneous reference system amounts to locally treating the inhomogeneous system as a metal — even when it is not. Conversely, the BA-LDA is bound to locally treat a metallic system as an insulator if the local occupation number is equal to 1 within the precison of the calculation. In both cases the problem arises because the local density (at one site only) is not enough to tell whether the system should be a metal or an insulator. The resulting *inverse band-gap problem* of the BA-LDA manifests itself as a possible lack of self-consistent metallic solutions when one of the local occupation numbers comes to within 5×10^{-3} to 1. For this reason there are some data points missing around $n = 1$ in Fig. 7. Luckily, such situations are rare: inhomogeneous Luttinger liquids with local occupation numbers very close to those for which the corresponding homogeneous system becomes a Mott insulator are realized only for some small regions in parameter space. Nevertheless, attempts to improve the BA-LDA functional in these regions, e.g., by smoothing out the discontinuity or by employing self-interaction corrections, are currently being made.

4. Summary and outlook

Density-functional theory provides a way to couch the many-body problem in terms of intensive density-like variables instead of wave functions (the Hohenberg-Kohn theorem), and a practical means of extracting observables from effective single-particle equations (the Kohn-Sham scheme). Both of these achievements have had a large impact on *ab initio* calculations, but the utility of DFT extends beyond any particular Hamiltonian: Many model Hamiltonians (in particular all that are expressed in terms of intensive density-like variables) can also be analysed with DFT. The formal proof of a Hohenberg-Kohn theorem and the setting up of a Kohn-Sham scheme are straightforward transcriptions of the corresponding *ab initio* procedures.

By contrast, the construction of suitable *xc* functionals is a more subtle matter, and crucially depends on the particular physics incorporated into the chosen model Hamiltonian [14]. In this work we have employed the Bethe Ansatz to develop an LDA-type functional for the one-dimensional Hubbard model. The resulting BA-LDA provides a very convenient and surprisingly accurate approach to large and inhomogeneous systems.

Clearly, the present work only represents a beginning. Some ways in which our functional can be improved have been mentioned at the end of Sec. 2 and of Sec. 3.4, respectively. Interesting applications of such functionals include the study of Friedel oscillations, determination of anomalous exponents, the calculation of finite-size effects, investigation of spin-density, charge-density, and bond-order waves [55], and the study of superlattices in the 1DHM. More generally, the extension of the present work to other model Hamiltonians (which require a different approach to the construction of functionals) will provide fertil ground for further applications of DFT. Work on the Heisenberg model is in progress [56].

The combination of computational efficiency with reasonable accuracy and applicability to large and inhomogeneous systems is the reason for the popularity of DFT in *ab initio* calculations in condensed-matter physics and quantum chemistry. Here we find that BA-based DFT for the 1DHM displays the same combination of features. We have thus reasons to hope that the BA-LDA may find fruitful applications in future studies of the 1DHM and other model Hamiltonians. Conversely, insights gained from such model calculations may also turn out to be useful for further development of *ab initio* DFT. The above findings about the energy gap in low-dimensional systems are one example [15], and the prospect of one day constructing an *ab initio* LDA for Luttinger liquids may be another.

Acknowledgments

This work was supported by FAPESP, CAPES, and CNPq.

References

[1] J. Hubbard, Proc. Roy. Soc. A **276**, 238 (1963). *ibid* **277**, 237 (1964). *ibid* **281**, 401 (1964).

[2] C. Herring in *Magnetism* Vol. IV, G. T. Rado and H. Suhl, eds. (Academic Press, New York, 1966).

[3] P. W. Anderson, Science **235**, 1196 (1987). *ibid* **256**, 1526 (1992).

[4] J. Voit, Rep. Prog. Phys. **58**, 977 (1995).

[5] P. Schlottmann, Int. J. Mod. Phys. B **11**, 355 (1997).

[6] N. F. Mott, *Metal-Insulator Transitions* 2nd. ed. (Taylor & Francis, London, 1990).

[7] F. Gebhard, *The Mott Metal-Insulator Transition*, Springer Tracts in Modern Physics Vol. 137 (Springer, New York, 1997).

[8] M. Imada, A. Fujimori, and Y. Tokura, Rev. Mod. Phys. **70**, 1039 (1998).

[9] M. Bockrath et al., Nature **397**, 598 (1999).

[10] H. W. Ch. Postma et al., Science **293**, 76 (2001).

[11] H. W. Ch. Postma et al., Phys. Rev. B **62**, 10653 (2000).

[12] O. M. Auslaender et al., Phys. Rev. Lett. **84**, 1764 (2000).

[13] M. Sassetti and B. Kramer, Phys. Rev. Lett. **80**, 1485 (1998).

[14] N. A. Lima, M. F. Silva, L. N. Oliveira, and K. Capelle, submitted (2001) [cond-mat 0112428].

[15] N. A. Lima, L. N. Oliveira, and K. Capelle, submitted (2002) [cond-mat 0205554].

[16] A. K. MacMahan, J. F. Annett, and R. M. Martin, Phys. Rev. B **42**, 6268 (1990).

[17] M. S. Hybertsen, M. Schlüter, and N. E. Christensen, Phys. Rev. B **39**, 9028 (1989).

[18] A. I. Liechtenstein, V. I. Anisimov, and J. Zaanen, Phys. Rev. B **52**, 5467 (1995).

[19] V. I. Anisimov, J. Zaanen, and O. K. Andersen, Phys. Rev. B **44**, 943 (1991).

[20] O. Gunnarsson and K. Schönhammer, Phys. Rev. Lett. **56**, 1968 (1986).

[21] K. Schönhammer and O. Gunnarsson, J. Phys. C **20**, 3675 (1987).

[22] K. Schönhammer, O. Gunnarsson, and R. M. Noack, Phys. Rev. B **52**, 2504 (1995).

[23] O. Gunnarsson and B. Lundqvist, Phys. Rev. B **13**, 4274 (1976).

[24] U. von Barth and L. Hedin, J. Phys. C **5**, 1629 (1972).

[25] D. M. Ceperley and B. J. Alder, Phys. Rev. Lett. **45**, 566 (1980).

[26] S. H. Vosko, L. Wilk, and M. Nusair, Can. J. Phys. **58**, 1200 (1980).

[27] J. P. Perdew and A. Zunger, Phys. Rev. B **23**, 5048 (1981).

[28] J. P. Perdew and Y. Wang, Phys. Rev. B **45**, 13244 (1993).

[29] F. D. M. Haldane, J. Phys. C **14**, 2585 (1981).

[30] H. A. Bethe, Z. Phys. **71** 205 (1931).

[31] E. H. Lieb and F. Y. Wu, Phys. Rev. Lett. **20**, 1445 (1968).

[32] E. Runge and G. Zwicknagl, Ann. der Physik **5**, 333 (1996).

[33] L. J. Sham and M. Schluter, Phys. Rev. Lett. **60**, 1582 (1988).

[34] O. Gunnarsson and K. Schönhammer, Phys. Rev. Lett. **60**, 1583 (1988).

[35] J. Carmelo and D. Baeriswyl, Phys. Rev. B **37**, 7541 (1988).

[36] A. A. Ovchinnikov, Sov. Phys. JETP **30**, 1160 (1970).

[37] L. N. Oliveira, E. K. U. Gross, and W. Kohn, Phys. Rev. Lett. **60**, 2430 (1988).

[38] P. Strange, *Relativistic Quantum Mechanics with Applications in Condensed Matter and Atomic Physics* (Cambridge University Press, Cambridge, 1998).

[39] A. M. Chang, L. N. Pfeiffer, and K. W. West, Phys. Rev. Lett. **77**, 2538 (1996).

[40] M. Grayson et al., Phys. Rev. Lett. **80**, 1062 (1998).

[41] A. Schwartz et al., Phys. Rev. B **58**, 1261 (1998).

[42] B. Dardel et al., Europhys. Lett. **24**, 687 (1993).

[43] J. D. Denlinger et al., Phys. Rev. Lett. **82**, 2540 (1999).

[44] C. Kim et al., Phys. Rev. Lett. **77**, 4054 (1996).

[45] P. Segovia et al., Nature **402**, 504 (1999).

[46] S. R. White, Phys. Rev. Lett. **69**, 2863 (1992). *ibid* **77**, 3633 (1993).

[47] G. Bedürftig, B. Brendel, H. Frahm, and R. M. Noack, Phys. Rev. B **58**, 10225 (1998).

[48] K. Capelle, L. N. Oliveira, Europhys. Lett. **49**, 376 (2000)

[49] K. Capelle, L. N. Oliveira, Phys. Rev. B **61**, 15228 (2000)

[50] M. F. Silva, K. Capelle, L. N. Oliveira, J. Mag. Magn. Mater. **226-230**, 1038 (2001)

[51] K. Capelle, M. F. Silva, L. N. Oliveira, J. Mag. Magn. Mater. **226-230**, 1017 (2001)

[52] J. P. Perdew, R. G. Parr, M. Levy, and J. L. Balduz, Phys. Rev. Lett. **49**, 1691 (1982).
[53] J. P. Perdew and M. Levy, Phys. Rev. Lett. **51**, 1884 (1983).
[54] L. J. Sham and M. Schlüter, Phys. Rev. Lett. **51**, 1888 (1983).
[55] N. A. Lima and L. N. Oliveira, unpublished.
[56] K. Capelle and V. L. Líbero, unpublished.

DEMONSTRATING THE EFFECTIVENESS OF A NONLOCAL DENSITY FUNCTIONAL DESCRIPTION OF EXCHANGE AND CORRELATION

Philip P. Rushton[1,2] and Stewart J. Clark[1]
[1]*Department of Physics, University of Durham, Science Laboratories, South Road, Durham DH1 3LE, UK.*
[2]*Department of Chemistry, University of Durham, Science Laboratories, South Road, Durham DH1 3LE, UK.*

Abstract The weighted density approximation (WDA) is a fully nonlocal exchange-correlation functional that depends only on the electron density $n(\mathbf{r})$. The WDA has received considerably less attention than other approaches such as the generalised gradient approximation GGA as it is computationally more expensive, but it does possess several advantages over conventional functionals. We demonstrate some of these features by considering various inhomogeneous electron gas systems in which the density is strongly inhomogeneous. Exchange-correlation energies, potentials and holes and pair-correlation functions are calculated and good agreement is found with recent variational Monte-Carlo (VMC) simulations. Comparisons are also made with the local density approximation (LDA) and GGA, and some shortcomings of these functionals are discussed.

1.   Introduction

The development of practical approximations to the unknown exchange-correlation (XC) functional in Kohn-Sham density functional theory (KS-DFT) [8, 11] remains a central and important task for the many-body description of electron interactions. One approximation that has received significant interest in recent years is the generalised gradient approximation (GGA) [13, 14], which builds upon the local density approximation (LDA) by including semilocal information, namely the reduced density gradient $s = |\nabla n(\mathbf{r})|/[2k_F n(\mathbf{r})]$. The GGA is not a unique functional form, consequently much research has been (and indeed still is) dedicated to fine-tuning GGA forms [21]. GGAs can be categorised broadly as non-empirical and (semi)-empirical. The non-empirical forms are derived from first principles and are guided by exact conditions of exchange and correlation [19, 22]. The development of non-empirical GGAs is considered to be complete, since optimal forms have been attained using this construction. Empirical GGAs are however still being actively pursued, mainly within the quantum chemistry community [2, 7]. These forms typically have several free parameters which are optimised by fitting to atomic and molecular data obtained from experiments.

An alternative avenue to explore are fully nonlocal functionals of the density such as the weighted density approximation (WDA). The WDA was created independently by Alonso and Girifalco WDA1, and Gunnarsson *et. al* WDA2 - well before the appearance of the GGA. However, due to the greater computational expense incurred by the nonlocal procedure compared with the GGA, the WDA has seldomly been used and remains relatively unknown. This is unfortunate since the WDA embodies many important features of the exact functional that are absent from commonly used approximations such as the LDA and the GGA. For instance, the WDA is in principle self-interaction free, it has the correct asymptotic limit for the XC-energy density, the asymptote of the potential has the correct form - differing from the exact result by a factor of 1/2, the sum rule on the XC-hole is fulfilled, it reduces to the LDA in the limit of a homogeneous electron gas, and it behaves correctly in low dimensional systems [3]. Other exact conditions can be built into the WDA through the construction of the model pair-correlation function, which is the only quantity approximated in the WDA [5].

The level of computational power available today means that the WDA is now a viable option. It is therefore necessary to establish the capabilities of the WDA, by performing tests on all manner of systems - from finite, (atoms,molecules) to infinite (solids) systems. The full range of systems can be examined using the plane-wave pseudopotential implementation of DFT. Fortunately the WDA is most efficiently realised within a reciprocal-space formalism [15, 9, 25], and so is perfectly suited to this method. However since the WDA is under-developed, WDA pseudopotentials have not yet been generated. Consequently, WDA calculations on real systems provide limited insight at present, since it is necessary to use the same exchange-correlation prescription for both core and valence electrons in order to assess the XC-functional properly [24]. One type of system that does not require pseudopotentials and yet can provide simple tests of an XC-functional is the inhomogeneous electron gas.

This work investigates a selection of model electron gas systems in which the density distribution varies strongly on the scale of the local Fermi wavelength λ_F. Several important exchange-correlation quantities such the total energy, energy density, potential, holes and pair-correlation functions will be examined using the WDA, and comparisons with the LDA and an established GGA will be made, wherever possible. The stimulus for this study was provided by a recent variational Monte-Carlo (VMC) investigation on an electron gas with a cosinusoidal perturbation applied in one dimension [17]. A remarkable result found in this work is that the point-wise difference between the LDA and the VMC energy density along the direction of inhomogeneity closely resembled the Laplacian of the density $\nabla^2 n(\mathbf{r})$. As demonstrated here, this striking feature is also exhibited by the WDA when the difference with respect to the LDA is taken. The WDA is also found to agree well with all other exchange-correlation quantities determined using the VMC method.

2. WDA theory

The WDA energy functional $E_{\rm XC}^{\rm WDA}[n(\mathbf{r})]$ is based on the exact expression given by the adiabatic connection method [12, 4],

$$E_{\rm XC}^{\rm WDA}[n(\mathbf{r})] = \frac{1}{2} \int n(\mathbf{r}) \, d\mathbf{r} \int \frac{n_{\rm XC}^{\rm WDA}(\mathbf{r}, \mathbf{r}')}{|\mathbf{r} - \mathbf{r}'|} \, d\mathbf{r}'. \tag{1}$$

The central approximation in the WDA is for the coupling-constant averaged pair-correlation function $g(\mathbf{r}, \mathbf{r}')$, which defines the exchange-correlation hole $n_{\rm XC}(\mathbf{r}, \mathbf{r}')$,

$$n_{\rm XC}(\mathbf{r}, \mathbf{r}') = n(\mathbf{r}') \left[g(\mathbf{r}, \mathbf{r}') - 1 \right]. \tag{2}$$

The WDA hole $n_{\rm XC}^{\rm WDA}(\mathbf{r}, \mathbf{r}')$ is usually written in the form,

$$n_{\rm XC}^{\rm WDA}(\mathbf{r}, \mathbf{r}') = n(\mathbf{r}') \, G^{\rm WDA}[\mathbf{r}, \mathbf{r}'; \tilde{n}(\mathbf{r})], \tag{3}$$

where $G^{\rm WDA}[\mathbf{r}, \mathbf{r}'; \tilde{n}(\mathbf{r})]$ is an analytic approximation that is chosen to satisfy exact limiting conditions. The WDA takes its name from the parameter $\tilde{n}(\mathbf{r})$ - that is the weighted density - which is a scalar field determined at each point in a system by satisfying the fundamental sum rule on the hole,

$$\int n_{\rm XC}^{\rm WDA}(\mathbf{r}, \mathbf{r}') \, d\mathbf{r}' = -1. \tag{4}$$

The XC-potential in the WDA is obtained in the usual manner by taking the functional derivative of $E_{\text{XC}}^{\text{WDA}}[n(\mathbf{r})]$. This leads to three terms,

$$v_{\text{XC}}^{\text{WDA}}(\mathbf{r}) = \frac{\delta E_{\text{XC}}^{\text{WDA}}[n(\mathbf{r})]}{\delta n(\mathbf{r})} = v_1(\mathbf{r}) + v_2(\mathbf{r}) + v_3(\mathbf{r}). \tag{5}$$

where the first term is simply the XC-energy density $\varepsilon_{\text{XC}}(\mathbf{r})$,

$$v_1(\mathbf{r}) = \frac{1}{2} \int n(\mathbf{r}') \frac{G^{\text{WDA}}[\mathbf{r},\mathbf{r}';\tilde{n}(\mathbf{r})]}{|\mathbf{r}-\mathbf{r}'|} d\mathbf{r}', \tag{6}$$

$v_2(\mathbf{r})$ arises from the fact that $G^{\text{WDA}}[\mathbf{r},\mathbf{r}';\tilde{n}(\mathbf{r}')]$ is asymmetric with respect to the interchange of particle coordintes, $\mathbf{r}$ and $\mathbf{r}'$,

$$v_2(\mathbf{r}) = \frac{1}{2} \int n(\mathbf{r}') \frac{G^{\text{WDA}}[\mathbf{r},\mathbf{r}';\tilde{n}(\mathbf{r}')]}{|\mathbf{r}-\mathbf{r}'|} d\mathbf{r}'. \tag{7}$$

Finally, $v_3(\mathbf{r})$ accounts for the dependence of $\tilde{n}(\mathbf{r})$ on the actual density, through $G^{\text{WDA}}[\mathbf{r},\mathbf{r}';\tilde{n}(\mathbf{r})]$,

$$v_3(\mathbf{r}) = \frac{1}{2} \int n(\mathbf{r}'') d\mathbf{r}'' \int \frac{n(\mathbf{r}')}{|\mathbf{r}'-\mathbf{r}''|} \times \frac{\delta G^{\text{WDA}}[\mathbf{r}',\mathbf{r}'';\tilde{n}(\mathbf{r}')]}{\delta n(\mathbf{r})} d\mathbf{r}'. \tag{8}$$

Gunnarsson and Jones [6] proposed a simple ansatz for $G^{\text{WDA}}[\mathbf{r},\mathbf{r}';\tilde{n}(\mathbf{r})]$ that contains two further parameters $\alpha(\mathbf{r})$ and $\beta(\mathbf{r})$, that are determined by satisfying two conditions in the limit of a homogeneous electron gas, namely the energy density and the sum rule. Consequently the WDA reduces to the LDA for a homogeneous electron gas. In this work we will employ the same type of ansatz as Gunnarsson and Jones, but choose the simplest possible analytic model - that of a Gaussian - so that $G^{\text{WDA}}[\mathbf{r},\mathbf{r}';\tilde{n}(\mathbf{r})]$ is written,

$$G^{\text{WDA}}[\mathbf{r},\mathbf{r}';\tilde{n}(\mathbf{r})] = \alpha(\mathbf{r}) e^{-\beta(\mathbf{r})|\mathbf{r}-\mathbf{r}'|^2}. \tag{9}$$

Details of our reciprocal-space implementation of the WDA within the CASTEP periodic code [28] are presented by Rushton *et. al* meSi.

3. Cosine-wave electron gas.

The details of this system are purposefully designed to follow the VMC work of Nekovee *et. al* as closely as possible. Here we expand on the results presented in referece [26]. Comparisons are made between the WDA and the VMC method and consequently the results of the VMC simulations obtained by Nekovee*et. al* nekovee should also be consulted.

Three model systems with strong density inhomogeneity are examined by applying an external potential of the form,

$$v_{\text{ext}}(\mathbf{r}) = v_q \cos(\mathbf{q} \cdot \mathbf{r}), \tag{10}$$

to an electron gas along one of the three dimensions. To be consistent with the VMC calculations we determine the densities from self-consistent Kohn-Sham calculations using the LDA. This procedure gives rise to density profiles that are approximately sinusoidal in one direction, with the modulation controlled by the amplitude v_q and the wavevector **q**. All three systems have the same average density given by,

$$n^\circ = \frac{3}{4\pi r_s^3}, \qquad (11)$$

with Wigner-Seitz radius $r_s = 2a_0$. We use cubic cells containing $N = 48$, 60 and 52 electrons. The size of the cells are designed such that they admit two, three and four periods of the cosine potential within the length of each cell, so that the density profiles contain the same number of peaks along the direction of inhomogeneity. The value of n° together with choice of N and the number of periods in the cell determines the wavevectors to be $q = 1.12\, k_F^\circ$, $1.56\, k_F^\circ$ and $2.18\, k_F^\circ$, where

$$k_F^\circ = (3\pi^2 n^\circ)^{\frac{1}{3}}, \qquad (12)$$

is the Fermi wavevector associated with the average density n°. For all three systems we use $v_q = 2.08\, \varepsilon_F^\circ$ where ε_F° is the Fermi energy corresponding to n°. The values of **q** and v_q used here lead to strong variations in the density on the scale of the local Fermi wavelength $\lambda_F^\circ = 2\pi/k_F^\circ$ and differ only slightly from the ones used in the VMC study.

3.1 XC-energy density $e_{xc}(\mathbf{r})$

In the VMC study, the XC-energy density was defined in a slightly different manner from the usual convention as,

$$e_{xc}^{VMC}(\mathbf{r}) = \frac{n(\mathbf{r})}{2} \int \frac{n_{xc}^{VMC}(\mathbf{r},\mathbf{r}')}{|\mathbf{r}-\mathbf{r}'|}\, d\mathbf{r}', \qquad (13)$$

that is, it is the integrand of the total energy expression. To be consistent with the VMC study, our WDA energy density $e_{xc}^{WDA}(\mathbf{r})$ for the three systems is similarly calculated as,

$$e_{xc}^{WDA}(\mathbf{r}) = n(\mathbf{r})\, \varepsilon_{xc}^{WDA}(\mathbf{r}) = \frac{n(\mathbf{r})}{2} \int \frac{n_{xc}^{WDA}(\mathbf{r},\mathbf{r}')}{|\mathbf{r}-\mathbf{r}'|}\, d\mathbf{r}'. \qquad (14)$$

The LDA energy density is similarly calculated as the integrand of the total energy,

$$e_{xc}^{LDA}(\mathbf{r}) = n(\mathbf{r})\, \varepsilon_{xc}^{hom}(\mathbf{r}). \qquad (15)$$

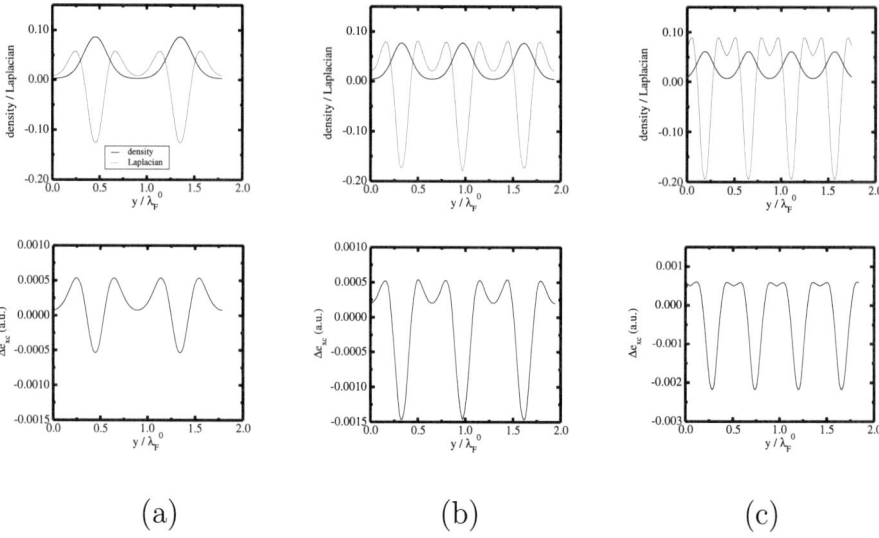

Figure 1. The upper panels display the density together with the Laplacian of the density $\nabla^2 n(\mathbf{r})$, and the lower panels show the exchange-correlation energy density difference $\Delta e_{\mathrm{XC}} = e_{\mathrm{XC}}^{\mathrm{LDA}} - e_{\mathrm{XC}}^{\mathrm{WDA}}$ for the cosine-wave systems with $q = 1.12\,k_F^0$ (a), $1.56\,k_F^0$ (b) and $2.18\,k_F^0$ (c).

Fig. 1 shows plots of $\Delta e_{\mathrm{XC}} = e_{\mathrm{XC}}^{\mathrm{LDA}} - e_{\mathrm{XC}}^{\mathrm{WDA}}$ calculated along the direction of inhomogeneity, as well as the corresponding density $n(\mathbf{r})$ and Laplacian of the density $\nabla^2 n(\mathbf{r})$ for the three systems. It is clear that Δe_{XC} bears a striking resemblance to $\nabla^2 n(\mathbf{r})$ - in complete accordance with the VMC findings. Although the LDA densities used here are different, the magnitude of the energy density deviations produced by the WDA are in very good agreement with the VMC results. Compare with Fig. 2 of the VMC study.

3.2 Total XC-energy

Table 1 shows the WDA total energy per electron $E_{\mathrm{XC}}^{\mathrm{WDA}}/N$, along with the difference relative to the LDA, $\Delta E_{\mathrm{XC}}^{\mathrm{LDA}}/N = (E_{\mathrm{XC}}^{\mathrm{LDA}} - E_{\mathrm{XC}}^{\mathrm{WDA}})/N$ and the GGA, $\Delta E_{\mathrm{XC}}^{\mathrm{GGA}}/N = (E_{\mathrm{XC}}^{\mathrm{GGA}} - E_{\mathrm{XC}}^{\mathrm{WDA}})/N$ for all three systems. The corresponding VMC values quoted in Ref. [17] are also given in brackets. Both the LDA and the GGA, yield positive deviations for the $q = 1.12 k_F^0$ system, but are negative for the other two systems. Interestingly, the LDA provides closer agreement with the WDA than the GGA for the $q = 1.56 k_F^0$ and $q = 2.18 k_F^0$ systems - the LDA gives deviations of -0.77%

Table 1. The WDA total energy $E_{\text{XC}}^{\text{WDA}}/N$, and the percentage differences relative to the LDA and the GGA, for the three cosine-wave systems. The values in brackets are the VMC values quoted from Table 1. of Ref. [17].

$q/k_{\text{F}}^{\text{o}}$		$E_{\text{XC}}^{\text{WDA}}/N$		$\Delta E_{\text{XC}}^{\text{LDA}}/N$		$\Delta E_{\text{XC}}^{\text{GGA}}/N$	
1.12	(1.11)	-0.3327	(-0.3289)	+1.47 %	(+1.28 %)	+0.30 %	(+0.03 %)
1.56	(1.55)	-0.3134	(-0.3127)	-0.77 %	(-0.16 %)	-2.46 %	(-2.37 %)
2.18	(2.17)	-0.2874	(-0.2882)	-3.34 %	(-2.29 %)	-5.57 %	(-4.86 %)

and -3.34%, compared with -2.46% and -5.57% for the GGA, for these two systems.

The agreement between the WDA and the VMC method cannot be rigourously quantified since the LDA densities used in this work are slightly different from those generated in the VMC study. It should also be noted that the VMC method contains sources of error [16] which although generally small, may alter the calculated results. Also, the actual LDA and GGA used in the VMC study are marginally different from those used here, since the homogeneous electron gas paramaters were re-optimised in both functionals using their VMC procedure, in order to eliminate finite size errors in their calculations. Nevertheless the agreement between the WDA and the VMC data is certainly promising. Exactly the same trends are observed for $\Delta E_{\text{XC}}^{\text{LDA}}/N$ and $\Delta E_{\text{XC}}^{\text{GGA}}/N$ for the two methods.

3.3   XC-potentials

Unfortunately, VMC XC-potentials cannot be compared with, and so only the three functionals will be examined. Fig. 2 shows the XC-potentials around a single density maximum for the $q = 1.12\,k_{\text{F}}^{\text{o}}$ and $q = 1.56\,k_{\text{F}}^{\text{o}}$ systems. The LDA and GGA potentials differ only slightly from each other, whereas the WDA decays more slowly, with appreciable differences occuring as the inhomogeneity increases. This is caused by the slower asymptotic behaviour $(-1/2r)$ for the WDA, originating from $v_1(\mathbf{r})$ in relation (6), compared with the exponential decay of the LDA and the GGA.

3.4   XC-holes

The WDA XC-hole $n_{\text{XC}}^{\text{WDA}}(\mathbf{r}, \mathbf{r}')$ for a reference electron at position $\mathbf{r}$ is determined from relations (3) and (9), using the LDA density as input. Since an explicit local hole can not be determined within the GGA formalism, only LDA, WDA and VMC holes are compared. The

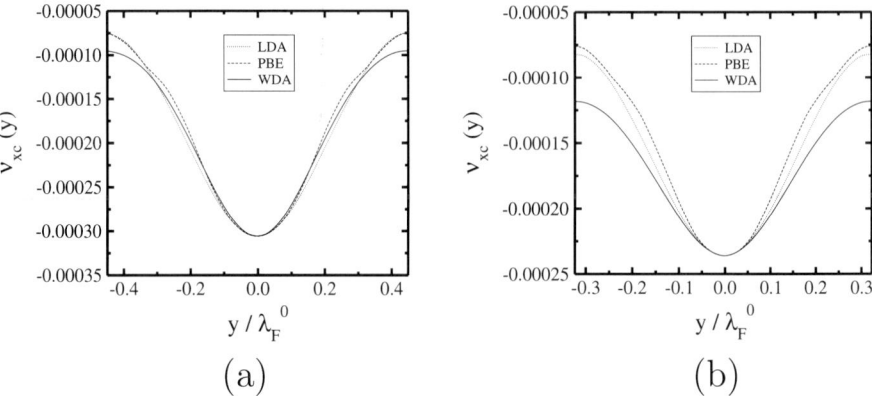

Figure 2. The exchange-correlation potential around a single density maximum in the $q = 1.12\,k_F^0$ and $q = 1.56\,k_F^0$ cosine-wave systems, for the three density functionals.

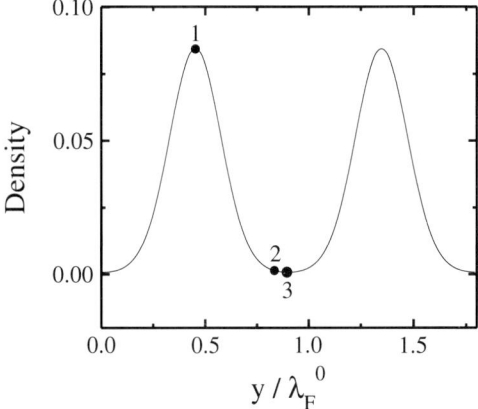

Figure 3. Diagram showing the positions of the electron along the $q = 1.12\,k_F^0$ density profile where the XC-holes in Fig. 4 are calculated. The locations are shown by filled circles and are labelled 1 (density maximum), 2 (near a density minimum) and 3 (density minimum).

LDA holes are generated using the parametrisation of the homogeneous gas pair-correlation function, derived by Perdew and Wang LDA$_n xc$.

The holes are plotted in a plane parallel to the direction of inhomogeneity, with the reference electron situated at various points along the direction of inhomogeneity. We consider three positions of interest - at a density maximum, near to the density minimum, and at a density minimum. These locations are labelled in Fig. 3.

When the electron is at a density maximum the WDA hole, shown in Fig. 4 (a), is centred directly at the site of the electron and is contracted in the direction of inhomogeneity. The LDA hole is in good agreement with the WDA - the on-top value $n_{\text{xc}}(\mathbf{r},\mathbf{r})$ in both cases are almost identical. However the LDA fails to describe the asymmetry of the hole due to its spherical construction. For positions away from the density maximum, the WDA hole becomes completely delocalised from the electron, which remains at the peak in the density, as shown in Fig. 4 (b). When the electron is at the density minimum, (Fig. 4 (c)) the hole-electron delocalisation effects become extremely prominent, since the WDA gives rise to a minimum in the hole either side of the reference electron. The LDA is clearly unable to imitate this nonlocality, as illustrated in Fig. 4 (c). In this region of low density, the LDA hole becomes shallow and very diffuse - expanding throughout space into neighbouring unit cells. This is a general feature of the LDA for relatively low densities, and is a direct consequence of satisfying the sum rule at such low values of the local density.

The WDA compares favourably with the VMC method, compare with Fig. 1 in the VMC study. At the density maximum the VMC hole displays the same anisotropy perpendicular to the direction of inhomogeneity, and has a very similar on-top value, (allowing for the different densities used). At a minimum in the density, the VMC method gives rise to a hole with the same characteristic nonlocal minima as the WDA.

The same general features are observed for the other two systems. Fig. 5 shows the WDA hole at a density maximum and minimum in the $q = 2.18\,k_{\text{F}}^{\text{o}}$ system.

3.5 Pair-correlation functions

The pair-correlation function, $g(\mathbf{r},\mathbf{r}')$, is highly anisotropic on a local scale for inhomogeneous systems. Again, a local GGA pair-correlation function cannot be determined, and the LDA pair-correlation function which can be calculated is spherically symmetric. Fig. 6 shows the WDA and LDA $g(\mathbf{r},\mathbf{r}')$ at three positions along the $q = 1.12\,k_{\text{F}}^{\text{o}}$ system - a density maximum (a), mid-way between a maximum and minimum (b) and at minimum (c). The anisotropy of the WDA results are very similar to the corresponding VMC data (not shown see Ref. [18]).

4. Large amplitude perturbations

Another way of increasing the inhomogeneity in ths system is to make the amplitude of the cosine perturbation larger. Therefore, for the last system we consider keeping the wavevector constant at $q = 3.22\,k_{\text{F}}^{\text{o}}$, and

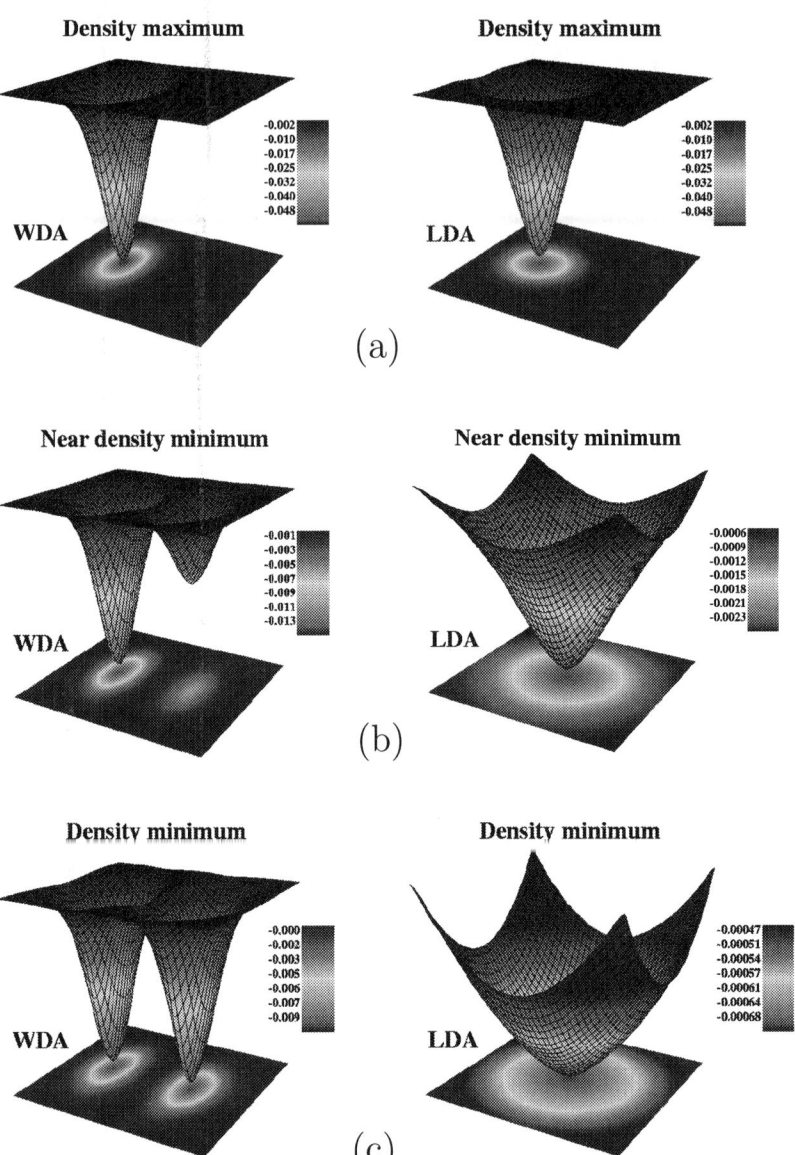

Figure 4. WDA and LDA exchange-correlation holes $n_{\text{XC}}(\mathbf{r},\mathbf{r}')$ calculated when a reference electron is located at a density maximum (a), near a minimum (b), and at a density minimum (c), in the $q = 1.12\,k_{\text{F}}^0$ system. The locations of the reference electron along the density profile are shown in Fig. 3 At the density minimum the electron (not shown) is situated in the centre of the plane, equidistant from the XC hole minima.

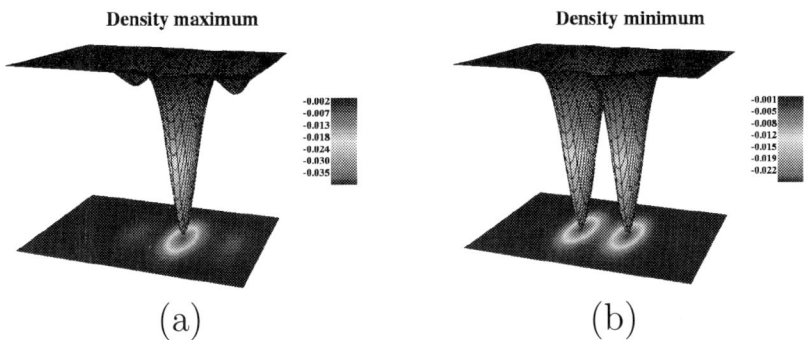

Figure 5. WDA exchange-correlation holes $n_{\text{XC}}(\mathbf{r}, \mathbf{r}')$, calculated for a reference electron located at a density maximum (a) and a density minimum (b), in the $q = 2.18\,k_{\text{F}}^0$ cosine-wave system.

examine three values for the amplitude, $v_q = 10\varepsilon_{\text{F}}^0, 100\,\varepsilon_{\text{F}}^0$ and $303\,\varepsilon_{\text{F}}^0$. The system has an average density $r_s = 4.31\,a_0$ with two density peaks along the direction of inhomogeneity. Fig. 7 shows the density and Laplacian of the density, together with the energy density difference between the LDA and WDA Δe_{xc}. It can be seen that the Laplacian looks less like Δe_{xc} than for the less inhomogeneous systems shown in Fig. 1. The positive contributions from $\nabla^2 n(\mathbf{r})$ are not displayed in Δe_{xc}, which become very large and negative around the density peaks. This results in greater deviations for the total energy, with respect to the WDA. For the three systems, the total energy differences are $8\%, 29\%$ and 34% for the LDA and $17\%, 41\%$ and 48% for the GGA, with the errors increasing with the size of the amplitude.

5. Conclusions

By examining several strongly inhomogeneous electron gas systems we have demonstrated that the WDA is a very promising exchange-correlation functional. The success of the WDA is attributed to its nonlocal description of the XC-hole. This was illustrated by several examples of highly anisotropic holes in the systems studied here. For the cosine-wave electron gas we showed that the WDA provides essentially the same results as the VMC method for all exchange-correlation quantities, but at a fraction of the computational cost. For each system the VMC simulations took more than 4000 CPU hours to perform [27], in contrast to the WDA, which on a standard desktop workstation requires only minutes for the systems detailed here.

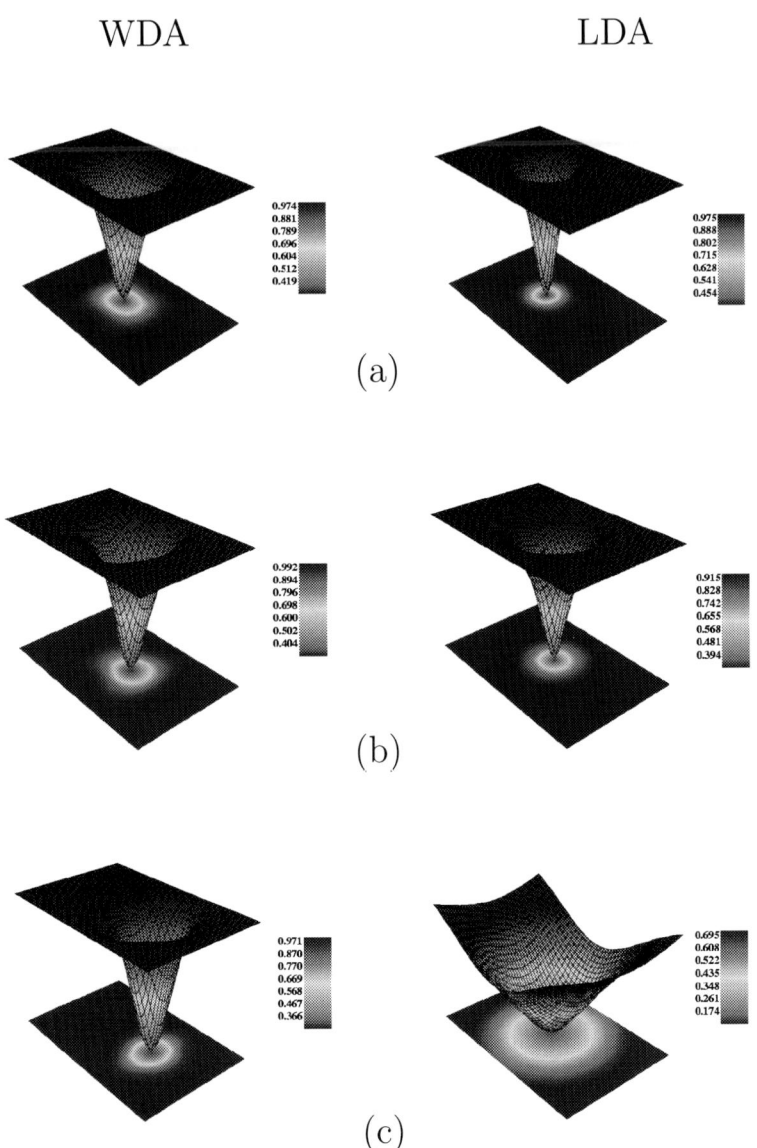

Figure 6. The pair-correlation function $g(\mathbf{r},\mathbf{r}')$ calculated for a reference electron located at a density maximum (a), mid-way between a maximum and minimum (b) and at a density minimum (c), in the $q = 1.12\, k_{\mathrm{F}}^0$ cosine-wave electron gas.

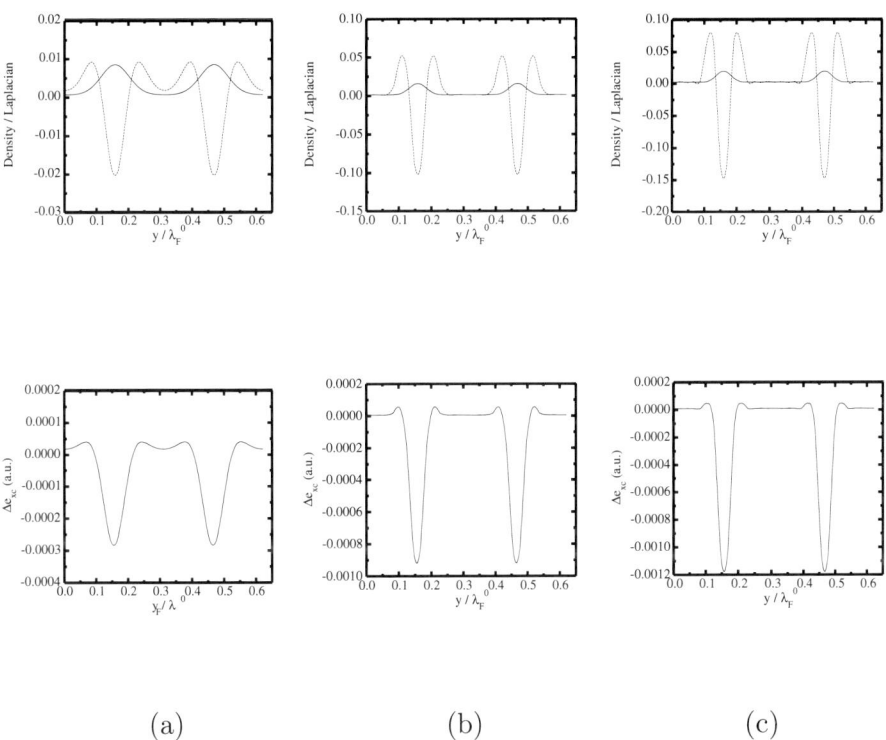

(a) (b) (c)

Figure 7. The upper panels display the density and the Laplacian of the density $\nabla^2 n(\mathbf{r})$, and the lower panels show the exchange-correlation energy density difference $\Delta e_{\mathrm{XC}} = e_{\mathrm{XC}}^{\mathrm{LDA}} - e_{\mathrm{XC}}^{\mathrm{WDA}}$, for the cosine-wave systems with constant wavevector $q = 3.22\, k_{\mathrm{F}}^0$ and increasing amplitudes, $v_q = 10\varepsilon_{\mathrm{F}}^0$ (a), $100\,\varepsilon_{\mathrm{F}}^0$ (b) and $303\,\varepsilon_{\mathrm{F}}^0$ (c).

The simplicity of the ansatz for G^{WDA}, together with the ability to calculate local exchange-correlation information effortlessly using the WDA (that cannot be calculated with semi-local functionals such as the GGA) and the promising results reported in this work, mean that the WDA is indeed a worthy functional to develop.

6. Acknowledgements

PPR thanks M. Nekovee for helpful discussions and acknowledges the EPSRC for financial support.

References

[1] Alonso, J. A. and L. A. Girifalco, Phys. Rev. B **17**, 3735 (1976).

[2] Becke, A. D. Phys. Rev. A **38**, 3098 (1988).

[3] Garcia-Gonzalez, P. Phys. Rev. B **62**, 2321 (2002).

[4] Gunnarsson, O and B. I. Lundqvist, Phys. Rev. B **13**, 4274 (1976).

[5] Gunnarsson, O., M. Jonson, and B. I. Lundqvist, Phys. Rev. B **20**, 3136 (1979); Solid State Commun. **24**, 765 (1977).

[6] Gunnarsson, O. and R. O. Jones, Phys. Scr. **21**, 394 (1980).

[7] Hamprecht, F. A., A. J. Cohen, D. J. Tozer, and N. C. Handy, J. Chem. Phys. **109**, 6264 (1998).

[8] Hohenberg, P. and W. Kohn, Phys. Rev. **136**, B864 (1964).

[9] Hybertsen, M. S. and S. G. Louie, Phys. Rev. B **30**, 5777 (1984).

[10] Jones, R. O. and O. Gunnarsson, Rev. Mod. Phys. **61** 689 (1989).

[11] Kohn, W. and L. J. Sham, Phys. Rev. **140**, A1133 (1965).

[12] Langreth, D. C. and J. P. Perdew, Solid State Commun. **17**, 1425 (1975).

[13] Langreth, D. C. and M. J. Mehl, Phys. Rev. Lett. **47**, 446 (1981).

[14] Langreth, D. C. and M. J. Mehl, Phys. Rev. B **28**, 1809 (1983).

[15] Marzari, N. and D. J. Singh, J. Phys. Chem. Solids **61**, 321 (2000).

[16] Nekovee, M., W. M. C. Foulkes, A. J. Williamson, G. Rajagopal, and R. J. Needs, Adv. Quantum Chem. **33** 189 (1999).

[17] Nekovee, M., W. M. C. Foulkes, and R. J. Needs, Phys. Rev. Lett. **87**, 036401 (2001).

[18] Nekovee, M., R. J. Needs and W. M. C. Foulkes (To be published).

[19] Perdew, J. P. *Electronic Structure of Solids '91*, edited by P. Ziesche and H. Eschrig (Akademie Verlag, Berlin, 1991); J. P. Perdew, J. A. Chevary, S. H. Vosko, K. A. Jackson, M. R. Pederson, D. J. Singh, and C. Fiolhais, Phys. Rev. B **46**, 6671 (1992).

[20] Perdew, J. P. and Y. Wang, Phys. Rev. B **46**, 12947 (1992).

[21] Perdew, J. P. and K. Burke, Int. J. Quan. Chem. **57**, 309 (1996).

[22] Perdew, J. P., K. Burke, and M. Ernzerhof, Phys. Rev. Lett. **77**, 3865 (1996).

[23] Pollack, L. and J. P. Perdew, J. Phys.: Condens. Matter **12** 1239 (2000).

[24] Rushton, P. P., S. J. Clark, and D. J. Tozer, Phys. Rev. B **63**, 115206 (2000).
[25] Rushton, P. P., D. J. Tozer, and S. J. Clark, Phys. Rev. B, **65**, 235203 (2002)(a).
[26] Rushton, P. P., D. J. Tozer, and S. J. Clark, Phys. Rev. B, **65**, 193106 (2002)(b).
[27] Rushton, P. P. Private communication with M. Nekovee. (2002)(c)
[28] Segall, M. D. Philip J. D. Lindan, M. J. Probert, C. J. Pickard, P. Hasnip, S. J. Clark and M. C. Payne, J. Phys.: Cond. Matter, **14**, 2717 (2002)

INCORPORATING THE VIRIAL FIELD INTO THE HARTREE-FOCK EQUATIONS

R.F.W. Bader
Department of Chemistry,
McMaster University,
Hamilton,
Ontario L8S 4M1, Canada

Abstract The underlying operational observation of chemistry - that of a functional group exhibiting a characteristic set of properties - requires that the density distribution of a group and hence its properties, be relatively insensitive to changes in its neighbouring groups. However, the fields that are used in the determination of the wave function and of the density in DFT are not short-range, but instead reflect the long-range nature associated with the individual e-n and e-e Coulombic fields. The observation upon which the theory of atoms in molecules is founded concerns the paralleling transferability of the electron density with all 'dressed' property densities. The virial field $V(\mathbf{r})$, the virial of the Ehrenfest force on an electron, describes the energy of interaction of an electron at some position $\mathbf{r}$ with all of the other particles in the system, averaged over the motions of the remaining electrons. When integrated over all space it yields the total potential energy of the molecule, including the nuclear energy of repulsion. Because of this inclusion, $V(\mathbf{r})$ yields the most short-range description possible of the potential interactions in a many-electron system. $V(\mathbf{r})$, as well as the kinetic energy density $G(\mathbf{r})$ and their sum $E_e(\mathbf{r})$, all demonstrably parallel the short-range behaviour underlying the transferable nature of $\rho(\mathbf{r})$. All three energy fields are determined by the one-matrix whose diagonal elements in addition determine $\rho(\mathbf{r})$. *Thus it is the one-matrix that is short-ranged and responsible for the observation of functional groups with characteristic, transferable properties in chemistry.* This paper describes a method for incorporating the virial field into the self-consistent field calculation to obtain an exact prescription of the 'average field' experienced by a single electron in a many- electron system.

1. Introduction

The self consistent potential in the Hartree-Fock equations consists of the electron-nuclear and electron-electron interactions. These Coulombic interactions are long-range. In describing the interactions that arise when atoms approach one another to form a molecule, a new interaction, the repulsion between the nuclei, is introduced. The interatomic electron-electron (e-e) and nuclear-nuclear (n-n) repulsions are both approximately one-half the magnitude of the inter-atomic electron-nuclear, (e-n) attractive interactions, the resulting difference between the repulsive and attractive interactions yielding the relatively small change in energy accompanying the approach of the atoms. Thus, because of the nuclear-nuclear contribution, the energy changes resulting from the formation of a molecule or from the relative vibrational displacements of its nuclei, are governed by a field that is relatively short-ranged compared to that determined by just the e-n and e-e interactions.

The virial field, the virial of the Ehrenfest force acting on an element of electron density, provides a spatial rendering of the total electrostatic potential field, one that includes the n-n, as well as the e-n and e-e interactions [1]. This a consequence of the virial of the e-n force yielding, in addition to $\hat{V}_{ne}$, the virial of the force that each nucleus with position vector $\mathbf{X}_\alpha$ exerts on the electron density. This contribution, when summed over all the nuclei, yields the negative of the virial of the forces acting on the nuclei together with the n-n repulsion energy, both of which appear in the statement of the molecular virial theorem given in equation (1)

$$-2T = V = \langle \hat{V}_{ne} \rangle + \langle \hat{V}_{ee} \rangle + \langle \hat{V}_{nn} \rangle + \sum_\alpha \mathbf{X}_\alpha \cdot \nabla_\alpha E \qquad (1)$$

V is the electronic virial and $-\nabla_\alpha E$ is the Hellmann-Feynman force acting on nucleus α. The electronic virial represents the electron's share of the total potential energy operator, a point made clear in terms of the electronic and nuclear sharing operators appearing in equation (2) [2]

$$\sum_i \left(-\mathbf{r}_i \cdot \nabla_i \hat{V} \right) + \sum_\alpha \left(-\mathbf{X}_\alpha \cdot \nabla_\alpha \hat{V} \right) = \sum_i \left(\mathbf{r}_i \cdot \hat{\mathbf{F}}_i \right) + \sum_\alpha \left(\mathbf{X}_\alpha \cdot \hat{\mathbf{F}}_\alpha \right) = \hat{V} \qquad (2)$$

Equation (2) is a consequence of the total potential energy operator $\hat{V}$ being a homogeneous function of degree -1 in the electronic and nuclear coordinates. It expresses $\hat{V}$ as a sum of virial operators $\mathbf{r}_K \cdot \hat{\mathbf{F}}_K$, and thus each determines the share of belonging to a given particle K. Thus the sum of the electronic virial operators defines the electron's share of the total potential energy as $\hat{V} - \sum_\alpha \mathbf{X}_\alpha \cdot \hat{\mathbf{F}}_\alpha$, a contribution that when averaged over the wave function, yields the virial V defined in equation (1).

Using the physics of an open system, the electronic virial can be expressed in terms of an energy density distributed in real space, the virial field $V(\mathbf{r})$ [1, 3]. This is the result of the physics of an open system defining for every measurable property of a system, a corresponding 'dressed' density distribution, one whose integration over an atomic basin yields the atom's additive contribution to that property. A dressed density distribution for some particular property accounts for the corresponding interaction of the density at some point in space with the remainder of the molecule [4]. The virial field $V(\mathbf{r})$ is a dressed density distribution of particular importance. It describes the energy of interaction of an electron at some position $\mathbf{r}$ with all of the other particles in the system, averaged over the motions of the remaining electrons. When integrated over all space it yields the total potential energy of the molecule, including the nuclear energy of repulsion and for a system in electrostatic equilibrium, it equals twice the molecule's total energy. The virial field condenses all of the electron-electron, electron- nuclear and nuclear-nuclear interactions described by the many- particle wave function into an energy density that is distributed in real space.

The expression for the virial field is obtained as the differential statement of the virial theorem for an open system, which is given in equation (2) for a system in a stationary state

$$2G(\mathbf{r}) = -\mathbf{r} \cdot \nabla \cdot \sigma(\mathbf{r}) + \nabla \cdot (\mathbf{r} \cdot \sigma(\mathbf{r})) - (\hbar^2/4m) \nabla^2 \rho(\mathbf{r}) \quad (3)$$

All quantities in equation (3) are defined in terms of the one- electron density matrix $\Gamma^{(1)}(\mathbf{r}, \mathbf{r}')$, the one-matrix; $G(\mathbf{r})$ is the positive definite kinetic energy density,

$$G(\mathbf{r}) = (\hbar^2/2m) \nabla \cdot \nabla' \Gamma^{(1)}(\mathbf{r}, \mathbf{r}') \Big|_{\mathbf{r}=\mathbf{r}'} \quad (4)$$

and $\sigma(\mathbf{r})$ is the quantum stress tensor, the negative of the momentum current density,

$$\sigma(\mathbf{r}) = (\hbar^2/4m) \{\nabla\nabla + \nabla'\nabla' - \nabla\nabla' - \nabla'\nabla\} \Gamma^{(1)}(\mathbf{r}, \mathbf{r}') \Big|_{\mathbf{r}=\mathbf{r}'} \quad (5)$$

The negative divergence of $\sigma(\mathbf{r})$ yields the Ehrenfest force acting on an element of the electron density,

$$\begin{aligned} \hat{\mathbf{F}}(\mathbf{r}) &= \int d\tau' \psi^* \left(-\nabla_r \hat{V}\right) \psi \\ &= -\nabla \cdot \sigma(\mathbf{r}) \end{aligned} \quad (6)$$

where the implied integration is over all spins and over the coordinates of all electrons save the one at **r** whose coordinates appear in the gradient of the full potential energy operator $\hat{V}$. The Ehrenfest force is thus a dressed density that describes the force exerted on the electron density at **r** by the nuclei and by the average distribution of the remaining electrons. The virial field, is defined in equation (3) and given by

$$V(\mathbf{r}) = -\mathbf{r} \cdot \nabla \cdot \sigma(\mathbf{r}) + \nabla \cdot (\mathbf{r} \cdot \sigma(\mathbf{r})) \qquad (7)$$

The integration of equation (3) over a proper open system yields, term for term, the virial theorem for an atom in a molecule, $-2T(\Omega) = V(\Omega)$, the final term in equation (3) vanishing because of the boundary condition of zero-flux in the gradient vector field of the density defining a proper open system. The atomic virial $V(\Omega)$ obtained by the integration of the virial field in equation (7) consists of two terms, the virial of the forces acting over the basin of the atom and a surface term giving the contribution from the virial of the Ehrenfest force acting on the atomic surface. The surface contribution vanishes for the total system with boundaries at infinity.

Among the most important results obtained from the quantum theory of atoms in molecules is its demonstration of the intimate relationship that exists between the distribution of the electron density within an atomic basin and the atom's contribution to a molecule's properties [1]. The reason for this is readily understood. Any two pieces of matter, including two atoms, are identical and possess identical properties only if they possess identical charge distributions, that is, they are indistinguishable in real space. Since an atom of theory is defined by its charge distribution as a bounded region of real space, its form necessarily reflects its properties. Consequently, each dressed density exhibits the same degree of atomic transferability between molecules as does the electron density. In particular, it was the observation of the simultaneous paralleling transferability of the electron density, the kinetic energy density and the virial field that led to the development of the theory of atoms in molecules [5]. Not only does $V(\mathbf{r})$ parallel the transferability of $\rho(\mathbf{r})$, its magnitude is found to be structurally homeomorphic with the electron density and thus every structure and change in structure exhibited by the topology of $\rho(\mathbf{r})$ is recovered by the topology of $|V(\mathbf{r})|$ [6]. Thus the electron density $\rho(\mathbf{r})$ appears as a locally scaled function of $V(\mathbf{r})$, the electronic potential energy density.

The underlying operational observation of chemistry - that of a functional group exhibiting a characteristic set of properties - requires that the density distribution of the group and hence its properties, be relatively insensitive to changes in its neighbouring groups. It has already been emphasized that $V(\mathbf{r})$ yields the most short-range description possible of the potential interactions in a many-electron system because it incorporates in a local manner the nuclear-nuclear force of repulsion. In the Hartree-Fock description of the formation of the H_2 molecule for example, the change in $\langle V_{en} \rangle = -1.3816$ a.u. The values of $\langle V_{ee} \rangle$ and $\langle V_{nn} \rangle$ are both approximately one-half the magnitude of $\langle V_{en} \rangle$, being $+0.6609$ and $+0.7207$ a.u. to yield a change in the virial of -0.266 a.u. which equals twice the change in the total energy $E = -0.133$ a.u. Clearly, without the contribution from the nuclear forces of repulsion, the magnitude of the SCF field is locally larger and of longer range. As a consequence, $V(\mathbf{r})$ as well as $G(\mathbf{r})$ and their sum $E_e(\mathbf{r})$, all parallel the short-range behaviour underlying the transferable nature of $\rho(\mathbf{r})$. All three energy fields are determined by the one-matrix whose diagonal elements in addition determine $\rho(\mathbf{r})$. *Thus it is the one-matrix that is short-ranged and responsible for the observation of functional groups with characteristic, transferable properties in chemistry.* [7]

2. Incorporation of V(r) into the Hartree-Fock Equations

The virial field is an exact prescription of the 'average field' experienced by a single electron in a many-electron system and the one that should be used in a self-consistent field calculation. It has been previously shown that expressing the potential energy of the electrons in terms of the virial field is necessary when one takes cognizance that forces of constraint are operative in determining the wave function for a molecule with a rigid nuclear framework [2]. That is, a molecule will be in mechanical equilibrium only if each of the forces acting on the nuclei is balanced by applied equal and opposite force. In classical mechanics, one obtains the equations of motion for a system subject to forces of constraint from an extended form of Hamilton's principle and a corresponding expression applies in the quantum case. In a classical system, the action integral is replaced by the variational integral I defined in equation (8),

$$I = \int dt \sum_i \left(p_i^2/2m + \mathbf{r}_i \cdot \hat{\mathbf{F}}_i \right) \tag{8}$$

and the variation is performed under the condition that the virtual work $\sum_i \hat{\mathbf{F}}_i \cdot \delta \mathbf{r}_i$ be consistent with the forces of constraint imposed on the system. The electronic energy E_e of a molecular system is expressed in an analogous manner with the potential energy defined in terms of the virial sharing operators, as in equation (9),

$$E_e = \sum_i \langle \hat{H}_i \rangle$$

$$= \left\langle \psi \left| \sum_i \left(\hat{t}_i + \mathbf{r}_i \cdot \hat{\mathbf{F}}_i \right) \right| \psi \right\rangle \quad (9)$$

It has been previously shown that the energy E_e yields the correct ground state energy if the variation of the energy integral includes the constraint that the forces on the nuclei equal pre-assigned values [2]. This procedure, as reviewed here, corresponds to the requirement that the virtual work of the forces of constraint acting on the nuclei vanish.

A molecule will be in mechanical equilibrium only if each of the nuclear forces $\langle \hat{\mathbf{F}}_\alpha \rangle = \langle \left(-\nabla_\alpha \hat{V} \right) \rangle = -\nabla_\alpha E$ is balanced by an applied force of equal and opposite magnitude, $-\nabla_\alpha E$. For a fixed nuclear configuration the condition of mechanical equilibrium for the n nuclei may be stated in terms of the virials of the nuclear and applied forces as

$$\left\langle \left(\mathbf{X}_\alpha \cdot \hat{\mathbf{F}}_\alpha + \mathbf{X}_\alpha \cdot \nabla_\alpha E \right) \right\rangle = 0, \quad \alpha = 1, 2, ..., n \quad (10)$$

and these conditions serve as the constraints to be imposed on the variation of the energy functional E_e. Since the force on each nucleus is constrained to a fixed value, the virtual work of the forces of constraint $\delta \langle \mathbf{X}_\alpha \cdot \hat{\mathbf{F}}_\alpha \rangle$ must vanish. This result coupled, with equation (2) – that the sum of the virial sharing operators for both electrons and nuclei equals the total potential energy operator $\hat{V}$– yields the desired result that the virtual work of the electrons will be constrained to equal the variation of the average potential energy of the molecule, that is,

$$\delta \left\langle \left(\sum_i \mathbf{r}_i \cdot \hat{\mathbf{F}}_i \right) \right\rangle = \delta \langle \hat{V} \rangle \quad (11)$$

The Hamiltonian in equation (9), termed the constrained Hamiltonian, is simply N times the contribution from one electron in both the classical and quantum cases. Hence the energy is expressed as a sum of contributions, one from each of the unconstrained particles. The virial sharing operator for a single electron is given by

$$\mathbf{r}_1 \cdot \hat{\mathbf{F}}_1 = -\mathbf{r}_1 \cdot \nabla_1 \hat{V}$$

$$= -\sum_\alpha Z_\alpha e^2 r_{1\alpha}^{-1} - \sum_\alpha \mathbf{X}_\alpha \cdot \nabla_1 \left(\sum_\alpha Z_\alpha e^2 r_{1\alpha}^{-1} \right) +$$

$$+ (N-1) \left\{ e^2 r_{12}^{-1} + r_2 \nabla_2 \left(e^2 r_{12}^{-1} \right) \right\} \quad (12)$$

where $\mathbf{r}_{1\alpha} = \mathbf{r}_1 - \mathbf{X}_\alpha$. The first two terms come from taking the electronic virial of the gradient of the electron-nuclear potential energy operator $\hat{V}_{en}$, as detailed in equation (13)

$$\begin{aligned}\mathbf{r}_1 \cdot \nabla_1 \hat{V}_{en} &= -\sum_\alpha Z_\alpha e^2 \mathbf{r}_1 \cdot \mathbf{r}_{1\alpha}/r_{1\alpha}^3 \\ &= -\sum_\alpha Z_\alpha e^2/r_{1\alpha} - \sum_\alpha Z_\alpha e^2 \mathbf{X}_\alpha \cdot \mathbf{r}_{1\alpha}/r_{1\alpha}^3 \end{aligned} \quad (13)$$

and it is this term that generates the virial of the force that the nuclei exert on the electron density, the force $\hat{\mathbf{F}}_{\alpha n} = -Z_\alpha e^2 \mathbf{r}_{1\alpha}/r_{1\alpha}^3$. When averaged over the state function, the virial sharing operator for a single electron yields [1]

$$\begin{aligned}\langle \psi | \mathbf{r}_1 \cdot \hat{\mathbf{F}}_1 | \psi \rangle &= \int d\mathbf{r}_1 \Big\{ \sum_\alpha \left(Z_\alpha e^2 r_{1\alpha}^{-1} - Z_\alpha e^2 \mathbf{X}_\alpha \cdot \mathbf{r}_{1\alpha}/r_{1\alpha}^3 \right) \rho(\mathbf{r}) + \\ &+ \int d\mathbf{r}_2 e^2 r_{12}^{-1} \Gamma^{(2)}(\mathbf{r}_1, \mathbf{r}_2) + \\ &+ \int d\mathbf{r}_2 \left(\mathbf{r}_2 \cdot \nabla_2 - \mathbf{r}_1 \cdot \nabla_1 \right) e^2 r_{12}^{-1} \Gamma^{(2)}(\mathbf{r}_1, \mathbf{r}_2) \Big\} \end{aligned} \quad (14)$$

where $\Gamma^{(2)}$ is the second-order density matrix normalized to $N(N-1)/2$. When the final integrations are performed, one recovers the terms for the virial V in equation (1). Integration over $\rho(\mathbf{r})$ yields $\langle \hat{V}_{ne} \rangle$ and the term $\sum_\alpha \mathbf{X}_\alpha \cdot \hat{\mathbf{F}}_{\alpha n} = \langle \hat{V}_{nn} \rangle + \sum_\alpha \mathbf{X}_\alpha \cdot \nabla_\alpha E$. Thus it is the electronic virial of the electron-nuclear force as detailed in equation (13) that yields a local description of the nuclear-nuclear repulsion in terms of the nuclear virial of the force each nucleus exerts on the density. The integration over $\Gamma^{(2)}$ in equation (14) yields $\langle \hat{V}_{ee} \rangle$, the final term in equation (14) vanishing when the integration is performed over all space. Setting the wave function in equation (9) equal to a single determinant of spin orbitals ϕ_i, the electronic energy E_e is given by

$$E_e = \sum_i \langle \phi_i | \hat{h}(i) + \sum_\alpha \mathbf{X}_\alpha \cdot \hat{\mathbf{F}}_{\alpha n} | \phi_i \rangle + \\ + \tfrac{1}{2} \sum_i \sum_j \langle \phi_i | \hat{J}_j - \hat{K}_j | \phi_i \rangle \quad (15)$$

where the one- and two-electron operators have their usual meaning and where $E_e = T + V = T + V + \sum_\alpha \mathbf{X}_\alpha \cdot \nabla_\alpha E$. Only when the forces on the nuclei vanish does E_e equal the Born-Oppenheimer energy for a rigid framework, $E = T + V$.

The energy E_e is to be minimized by varying the orbitals subject to the constraints given in equation (10). This can be done using the constrained variation method developed for the Hartree- Fock procedure by Mukherji and Karplus [8], wherein the coefficients of the basis functions are varied to satisfy the constraints to some pre-assigned level of agreement. The requirement of satisfying a pre-assigned set of Hellmann-Fenyman forces on the nuclei can be effectively met by employing a basis set denoted by $(r, r', r'', ...)$ which includes for each function χ_r the function $\chi_{r'}$, its derivative with respect to $\mathbf{X}_\alpha$, the coordinate of the nucleus on which it is centred [9, 10]. The use of such a basis has been shown to satisfy the Hellmann-Feynman theorem to a satisfactory degree of accuracy using only a few derivatives.

The Hartree-Fock equation for each spin orbital is modified by the inclusion of the virial of the forces exerted on the electron density by the nuclei, as shown in equation (16)

$$\left(\hat{h}(i) + \sum_\alpha \mathbf{X}_\alpha \cdot \hat{\mathbf{F}}_{\alpha n}\right)|\phi_i\rangle + \sum_j \left(\hat{J}_j - \hat{K}_j\right)|\phi_i\rangle = v_i |\phi_i\rangle \quad (16)$$

The orbital energy v_i will contain the orbital's share of the nuclear repulsion along with the virial of any net forces acting on the nuclei, a portion of the term $\langle \hat{V}_{nn} \rangle + \sum_\alpha \mathbf{X}_\alpha \cdot \nabla_\alpha E$. Since the field determining each spin orbital differs from the Hartree-Fock case by the inclusion of the virial of the electric field that each nucleus exerts on the density, one anticipates that the 'virial' orbitals as well as their energies, will differ from the corresponding Hartree-Fock quantities, the sum of the orbital energies yielding

$$\sum_i v_i = E_e + \langle \hat{V}_{ee} \rangle$$
$$= E + \sum_\alpha \mathbf{X}_\alpha \cdot \nabla_\alpha E + \langle \hat{V}_{ee} \rangle \quad (17)$$

The energy E is the Born-Oppenheimer energy for a fixed nuclear geometry and hence $E = E_{HF} + \langle \hat{V}_{nn} \rangle$ and $E_e = E + \sum_\alpha \mathbf{X}_\alpha \cdot \nabla_\alpha E$.

Will the energy E_e determined in equation (15) by the inclusion of the virial field in the Hartree-Fock equations as described in equation (16), differ from the energy $E = E_{HF} + \langle \hat{V}_{nn} \rangle + \sum_\alpha \mathbf{X}_\alpha \cdot \nabla_\alpha E$ that is obtained using the same basis set in solving the usual Hartree-Fock equations? Work is in progress to answer this question.

References

[1] R.F.W. Bader, Atoms in Molecules: a Quantum Theory (Oxford University Press, Oxford UK, 1990).

[2] S. Srebrenik, R.F.W. Bader, and T.T. Nguyen-Dang, J. Chem. Phys. 68, 3667 (1978).

[3] R.F.W. Bader, Phys. Rev. B 49, 13348 (1994).

[4] R.F.W. Bader, Can. J. Chem. 76, 973 (1998).

[5] R.F.W. Bader and P.M. Beddall, J. Chem. Phys. 56, 3320 (1972).

[6] T.A. Keith, R.F.W. Bader, and Y. Aray, Int. J. Quantum Chem. 57, 183 (1996).

[7] R.F.W. Bader, Int. J. Quantum Chem. 56, 409 (1995).

[8] A. Mukherji and M. Karplus, J. Chem. Phys. 38, 44 (1963).

[9] H. Nakatsuji, K. Kanda, and T. Yonezawa, Chem. Phys. Lett. 75, 340 (1980).

[10] H. Nakatsuji, T. Hayakawa, and M. Hada, Chem. Phys. Lett. 80, 94 (1981).

HOHENBERG-KOHN THEOREM AND CONSTRAINED SEARCH FORMULATION FOR DIAGONAL SPIN DENSITY FUNCTIONAL THEORY

Nikitas I. Gidopoulos
ISIS Facility, Rutherford Appleton Laboratory
Chilton, Didcot, Oxon, OX11 0QX, England, UK

Abstract The foundations of Spin-Density-Functional Theory are reviewed briefly. In the general case, the mapping between spin-potentials and spin-densities has not been proven to be invertible yet. It is shown that when we restrict attention to diagonal spin-potentials, the mapping between potentials and densities becomes invertible and Diagonal-Spin-Density-Functional Theory is seen to share a similar rigourous foundation to spinless Density Functional Theory.

1. Introduction

Density-Functional Theory (DFT) has long been established as a popular method to study the electronic structure in physics and chemistry. The success is largely due to the ease of calculation, i.e. the solution, under the Kohn-Sham (KS) scheme [1], of single-particle Hartree-like equations, but equally so, the success is owed to the rigorous foundation of the theory, based on the Hohenberg Kohn (HK) theorem [2], or the equivalent constrained search formulation [3, 4].

A large number of DFT applications in fact employ Spin-Density-Functional Theory (SDFT) which shares the computational simplicity of DFT but not its rigorous foundations. As we shall see, a HK theorem has not been proven for general SDFT that would establish the invertibility of the mapping between spin-potentials and spin-densities. On the contrary, von Barth and Hedin demonstrated that at least in the one-electron case, many different spin-potentials may share a common ground state.

In spite of this, Perdew and Zunger proposed a formulation of SDFT based on the constrained search approach [5]. Recently however, Capelle and Vignale [6] demonstrated using several counterexamples, that spin-potentials in Diagonal-Spin-Density-Functional Theory (DSDFT) are not uniquely determined and the functional derivative of density functionals are not well defined.

In this work we show that the HK theorem holds in DSDFT, provided that diagonal spin-potentials are considered different, if they differ by more than a spin-constant. We also reformulate the constrained search scheme to respect this basic property.

2. Spin-Density-Fuctional Theory

DFT was extended to include spin-dependent potentials by von Barth and Hedin [7] and independently by Pant and Rajagopal [8].

A spin-dependent potential $\mathcal{V}$, can be written in either of the two equivalent forms:

$$\mathcal{V} = (V^{\uparrow\uparrow}, V^{\uparrow\downarrow}, V^{\downarrow\uparrow}, V^{\downarrow\downarrow}) = (V; \mu_0 \mathbf{B}) \tag{1}$$

where, $\mu_0 = e\hbar/(2mc)$ and $-e$ the electronic charge. The two forms are related by:

$$V(\mathbf{r}) = \frac{V^{\uparrow\uparrow}(\mathbf{r}) + V^{\downarrow\downarrow}(\mathbf{r})}{2} \tag{2}$$

$$\mu_0 B_x(\mathbf{r}) = \frac{V^{\uparrow\downarrow}(\mathbf{r}) + V^{\downarrow\uparrow}(\mathbf{r})}{2} \tag{3}$$

$$\mu_0 B_y(\mathbf{r}) = \frac{V^{\uparrow\downarrow}(\mathbf{r}) - V^{\downarrow\uparrow}(\mathbf{r})}{-2i} \tag{4}$$

$$\mu_0 B_z(\mathbf{r}) = \frac{V^{\uparrow\uparrow}(\mathbf{r}) - V^{\downarrow\downarrow}(\mathbf{r})}{2} \tag{5}$$

SDFT deals with systems described by a Hamiltonian $\hat{H}_\mathcal{V}$ of the form (in second quantization):

$$\hat{H}_\mathcal{V} = \hat{T} + \hat{W} + \hat{\mathcal{V}}$$

where,

$$\hat{T} = -\frac{\hbar^2}{2m} \sum_{\alpha=\uparrow,\downarrow} \int d\mathbf{r} \left(\hat{\psi}^\alpha(\mathbf{r})\right)^\dagger \nabla^2 \hat{\psi}^\alpha(\mathbf{r}) \qquad (7)$$

$$\hat{W} = \frac{e^2}{2} \sum_{\alpha,\beta=\uparrow,\downarrow} \int\int \frac{d\mathbf{r}d\mathbf{r}'}{|\mathbf{r}-\mathbf{r}'|} \left(\hat{\psi}^\alpha(\mathbf{r})\right)^\dagger \hat{\rho}^{\beta\beta}(\mathbf{r}') \hat{\psi}^\alpha(\mathbf{r}) \qquad (8)$$

$$\hat{V} = \sum_{\alpha,\beta=\uparrow,\downarrow} \int d\mathbf{r} \left(\hat{\psi}^\alpha(\mathbf{r})\right)^\dagger V^{\alpha\beta}(\mathbf{r}) \hat{\psi}^\beta(\mathbf{r})$$

$$= \int d\mathbf{r}\, V(\mathbf{r}) \sum_{\alpha=\uparrow,\downarrow} \hat{\rho}^{\alpha\alpha}(\mathbf{r}) + \int d\mathbf{r}\, \mu_0 \mathbf{B}(\mathbf{r}) \cdot \hat{\mathbf{m}}(\mathbf{r}) \qquad (9)$$

where, $\hat{\rho}^{\alpha\beta}(\mathbf{r}) = \left(\hat{\psi}^\alpha(\mathbf{r})\right)^\dagger \hat{\psi}^\beta(\mathbf{r})$ and $\hat{\mathbf{m}}(\mathbf{r})$ the magnetic moment operator

$$\hat{\mathbf{m}}(\mathbf{r}) = \sum_{\alpha\beta=\uparrow,\downarrow} \int d\mathbf{r} \left(\hat{\psi}^\alpha(\mathbf{r})\right)^\dagger \sigma^{\alpha\beta} \hat{\psi}^\beta(\mathbf{r}) \qquad (10)$$

or, explicitly, in terms of the components, $\hat{m}_x(\mathbf{r}) = \hat{\rho}^{\uparrow\downarrow}(\mathbf{r}) + \hat{\rho}^{\downarrow\uparrow}(\mathbf{r})$, $i\hat{m}_y(\mathbf{r}) = \hat{\rho}^{\uparrow\downarrow}(\mathbf{r}) - \hat{\rho}^{\downarrow\uparrow}(\mathbf{r})$ and $\hat{m}_z(\mathbf{r}) = \hat{\rho}^{\uparrow\uparrow}(\mathbf{r}) - \hat{\rho}^{\downarrow\downarrow}(\mathbf{r})$. Finally, σ is the vector of the 2×2 Pauli spin-matrices $\sigma = (\sigma_x, \sigma_y, \sigma_y)$.

Note that the Hamiltonian does not describe fully systems of electrons in magnetic fields, because the vector potential is absent from the kinetic energy operator. As it stands, the Hamiltonian is appropriate to describe approximately, weak magnetic fields, or cases where spontaneous magnetization occurs in the absence of a magentic field.

Solving the Schrödinger equation for the ground state, one obtains $\Psi_\mathcal{V}$ with spin density $\varrho = (\rho^{\uparrow\uparrow}, \rho^{\uparrow\downarrow}, \rho^{\downarrow\uparrow}, \rho^{\downarrow\downarrow})$, where, $\rho^{\alpha\beta}(\mathbf{r}) = \langle \Psi_\mathcal{V}|\hat{\rho}^{\alpha\beta}(\mathbf{r})|\Psi_\mathcal{V}\rangle$.

2.1 No Hohenberg-Kohn Theorem in general SDFT

The invertibility of the mapping between spin-potentials and densities has not been proven in SDFT. In order to establish this 1-1 mapping between potentials and densities, one would have to show that two other mappings are separately invertible: First one should prove that different spin potentials $\mathcal{V}$ and $\mathcal{V}'$ always lead to different ground states $\Psi_\mathcal{V}$ and $\Psi_{\mathcal{V}'}$. Secondly that different ground states, arising from different potentials always lead to different spin densities ϱ and ϱ'. This second step, which establishes that the mapping between ground states and densities is invertible was straightforward to show [9]: Consider two states $\Psi_\mathcal{V}$ and $\Psi_{\mathcal{V}'}$ which are ground states of two different potentials. Can the spin densities be the same? Using the Rayleigh-Ritz minimum principle we have

$$\langle \Psi_V | H_V | \Psi_V \rangle < \langle \Psi_{V'} | H_V | \Psi_{V'} \rangle, \tag{11}$$

$$\langle \Psi_{V'} | H_{V'} | \Psi_{V'} \rangle < \langle \Psi_V | H_{V'} | \Psi_V \rangle, \tag{12}$$

Subtracting the two inequalities we obtain:

$$\sum_{\alpha,\beta=\uparrow,\downarrow} \int d^3\mathbf{r} \, [V'^{\alpha\beta}(\mathbf{r}) - V^{\alpha\beta}(\mathbf{r})] \, [\rho'^{\alpha\beta}(\mathbf{r}) - \rho_{\alpha\beta}(\mathbf{r})] < 0. \tag{13}$$

Assuming the opposite $\varrho = \varrho'$, we get $0 < 0$, which is absurd.

The first mapping however, between spin-potentials and wavefunctions has never been shown to be invertible. On the contrary, von Barth and Hedin [7], used a counterexample to demonstrate that with any one-electron Hamiltonian $\hat{H}_V^{\alpha\beta} = -\frac{1}{2}\delta^{\alpha\beta}\nabla^2 + V^{\alpha\beta}(\mathbf{r})$, there is a whole class of different one-electron Hamiltonians with the same ground state. Consider the ground state $\begin{pmatrix} \psi^\uparrow(\mathbf{r}) \\ \psi^\downarrow(\mathbf{r}) \end{pmatrix}$ of H_V. Then the spin potential $(v(\mathbf{r}); \mathbf{b}(\mathbf{r}))$ with $v(\mathbf{r})$ arbitrary and

$$b_x(\mathbf{r}) = -v(\mathbf{r}) \frac{(\psi^\uparrow(\mathbf{r}))^* \psi^\downarrow(\mathbf{r}) + \psi^\uparrow(\mathbf{r}) (\psi^\downarrow(\mathbf{r}))^*}{|\psi^\uparrow(\mathbf{r})|^2 + |\psi^\downarrow(\mathbf{r})|^2} \tag{14}$$

$$b_y(\mathbf{r}) = i\,v(\mathbf{r}) \frac{(\psi^\uparrow(\mathbf{r}))^* \psi^\downarrow(\mathbf{r}) - \psi^\uparrow(\mathbf{r}) (\psi^\downarrow(\mathbf{r}))^*}{|\psi^\uparrow(\mathbf{r})|^2 + |\psi^\downarrow(\mathbf{r})|^2} \tag{15}$$

$$b_z(\mathbf{r}) = -v(\mathbf{r}) \frac{|\psi^\uparrow(\mathbf{r})|^2 - |\psi^\downarrow(\mathbf{r})|^2}{|\psi^\uparrow(\mathbf{r})|^2 + |\psi^\downarrow(\mathbf{r})|^2} \tag{16}$$

satisfies $\sum_\beta \left[v(\mathbf{r}) \delta^{\alpha\beta} + \mathbf{b}(\mathbf{r}) \cdot \sigma^{\alpha\beta} \right] \psi^\beta(\mathbf{r}) = 0$, and therefore any one-electron Hamiltonian $\hat{H}_{V+(v;\mathbf{b})}$ with v arbitrary but small enough, has the same ground state as $\hat{H}_V$.

The question whether different spin-potentials may have a common ground state in the case of more than one electrons is important and has received increased attention recently [6, 10, 11].

It is helpful at this point to distinguish between the case of diagonal potentials (when $V^{\alpha\beta} = 0$, for $\alpha \neq \beta$) or axial magnetic fields $\mathbf{B}(\mathbf{r}) = B_z(\mathbf{r})\,\mathbf{z}$ (the direction of the magnetic field defining the z-axis), and the general case when $V^{\alpha\beta}$ or B_x, B_y need not vanish.

In the simplest case of diagonal spin-potentials, or axial magnetic fields, we will see that unless the ground state is completely polarized, spin-potentials which differ by more than a spin-dependent constant always have different ground states. First to observe this using a different proof, were Eschrig and Pickett in Ref. [10]. Hence, the invertibility of the mapping between diagonal-spin-potentials and densities can be established, as long as we define that two spin-potentials are different, when they differ by more than a spin-dependent constant. Note that the counterexample of von Barth and Hedin, requires non-zero magnetic field along the x and/or y axes and is not applicable.

Eschrig and Pickett extended their analysis to the general case and discovered that if two spin-potentials had a common eigenstate, then their difference could only take the form:

$$\delta \mathcal{V} = \int d\mathbf{r}\; \delta \mathbf{B}(\mathbf{r}) \cdot \hat{\mathbf{m}}(\mathbf{r}),\; with\; |\delta \mathbf{B}(\mathbf{r})|^2 = 1 \qquad (17)$$

The important question whether two general spin-potentials may indeed have a common many-electron ground state remains open.

2.2 Constrained search formulation of SDFT

The Rayleigh-Ritz minimum principle seemed sufficient to formulate SDFT even without the proof of a HK theorem. Perdew and Zunger [5] proposed the following constrained search formulation for SDFT: The ground state energy of $\hat{H}_\mathcal{V}$ is given by

$$E_\mathcal{V} = \min_{\Psi \to N} \langle \Psi | \hat{H}_\mathcal{V} | \Psi \rangle. \qquad (18)$$

As usual, search for the minimum in two steps. First minimize the expectation value of $H_\mathcal{V}$ over all normalized N-particle states giving a fixed spin-density $\varrho = (\rho^{\uparrow\uparrow}, \rho^{\uparrow\downarrow}, \rho^{\downarrow\uparrow}, \rho^{\downarrow\downarrow})$:

$$E_\mathcal{V}[\varrho] = F[\varrho] + \sum_{\alpha,\beta=\uparrow,\downarrow} \int d^3\mathbf{r}\; V^{\alpha\beta}(\mathbf{r}) \rho^{\alpha\beta}(\mathbf{r}) \qquad (19)$$

where, the internal energy functional is:

$$F[\varrho] = \min_{\Psi \to \varrho} \langle \Psi | \hat{T} + \hat{W} | \Psi \rangle. \qquad (20)$$

In order to fix the spin-density ϱ: $\langle \Psi | (\psi^\alpha(\mathbf{r}))^\dagger \psi^\beta(\mathbf{r}) | \Psi \rangle$, one must introduce the Lagragian multipliers $\lambda^{\alpha\beta}[\varrho](\mathbf{r})$. One must vary the quantity

$$\langle \Psi | \left[\hat{T} + \hat{W} + \sum_{\alpha,\beta=\uparrow,\downarrow} \int d\mathbf{r}\; \left(\hat{\psi}^\alpha(\mathbf{r})\right)^\dagger \lambda^{\alpha\beta}(\mathbf{r}) \hat{\psi}^\beta(\mathbf{r}) \right] | \Psi \rangle \qquad (21)$$

The minimizing state will satisfy:

$$\left[\hat{T} + \hat{W} + \sum_{\alpha,\beta=\uparrow,\downarrow} \int d\mathbf{r} \left(\hat{\psi}^\alpha(\mathbf{r})\right)^\dagger \lambda^{\alpha\beta}[\varrho](\mathbf{r})\hat{\psi}^\beta(\mathbf{r})\right] \Psi[\varrho] = E[\varrho]\,\Psi[\varrho] \tag{22}$$

The Lagrangian multipliers play the role of the external spin-dependent potential, whose ground state density is ϱ. However, in the absence of a HK theorem, one cannot ascertain that when ϱ coincides with the ground state density of $\mathcal{V}$ then $\lambda[\varrho]$ must be the same as $\mathcal{V}$.

Clearly, $E_\mathcal{V}[\varrho] \geq E_\mathcal{V}$ and in order to obtain the ground state energy, one has to minimize $E_\mathcal{V}[\varrho]$ over all spin-densities ϱ integrating to N particles:

$$E_\mathcal{V} = \min_{\varrho \to N} E_\mathcal{V}[\varrho]$$

where the constraint is,

$$\int d^3\mathbf{r}\left(\rho^{\uparrow\uparrow}(\mathbf{r}) + \rho^{\downarrow\downarrow}(\mathbf{r})\right) = N \tag{23}$$

The minimum is attained at the correct ground state density, $E_\mathcal{V} = E_\mathcal{V}[\varrho_\mathcal{V}]$. For that density and assuming that the partial functional derivatives of $E_\mathcal{V}[\varrho]$ with respect to the components of ϱ at the exact spin-density exist, we will have:

$$\frac{\partial}{\partial \rho^{\alpha\beta}(\mathbf{r})}\left[E_\mathcal{V}[\varrho] - \mu \int d^3\mathbf{r}\left(\rho^{\uparrow\uparrow}(\mathbf{r}) + \rho^{\downarrow\downarrow}(\mathbf{r})\right)\right]\bigg|_{\varrho=\varrho_\mathcal{V}} = 0 \tag{24}$$

leading to:

$$V^{\alpha\beta}(\mathbf{r}) = -\frac{\partial F[\varrho]}{\partial \rho^{\alpha\beta}(\mathbf{r})}\bigg|_{\varrho_\mathcal{V}} + \mu\,\delta_{\alpha\beta} \tag{25}$$

The chemical potential $\mu = \mu(N)$ appears as the Lagrange multiplier to satisfy the constraint (23) that the density integrates to N particles.

The assumption that the partial functional derivatives of $E_\mathcal{V}[\varrho]$ with respect to $\rho^{\alpha\beta}$ exist, and therefore are unique [6], is equivalent or stronger to the assumption that the HK theorem must hold, something which has not been shown in general. In the one-electron case for example, we expect that the assumption of differentiability must fail. In the many-electron case, Capelle and Vignale [6] demonstrated recently that the functional derivative cannot exist, using several counterexamples from Diagonal-Spin-Density-Functional-Theory (SDFT).

3. Diagonal-Spin-Density-Functional theory

Often the original potential $\mathcal{V}$ is diagonal $V^{\uparrow\downarrow} = V^{\downarrow\uparrow} = 0$, or $B_x = B_y = 0$, in which case we will use the notation $V^{\uparrow} = V^{\uparrow\uparrow}$, $V^{\downarrow} = V^{\downarrow\downarrow}$, or $B_z = B$ and $\mathcal{V} = (V^{\uparrow}, V^{\downarrow})$, or, $\mathcal{V} = (V; B)$.

Restricting to such diagonal potentials, we may resort to a simpler formulation of SDFT which involves only the spin up and down components of the density $\varrho = (\rho^{\uparrow}, \rho^{\downarrow})$, where $\rho^{\uparrow} = \rho^{\uparrow\uparrow}$, $\rho^{\downarrow} = \rho^{\downarrow\downarrow}$, as conjugate to the spin up and down potentials $V^{\uparrow}$ and $V^{\downarrow}$. In DSDFT, the mapping between spin-potentials and ground state wavefunctions becomes invertible, provided one treats two spin-potentials as different if they differ by more than just a spin-constant [10, 11].

The key to the proof of invertibility is to note that a constant shift in $B(\mathbf{r})$ does not alter the eigenstates of $\hat{H}_{(V;B)}$, although it may change their order by causing the levels to cross. Hence, the ground state may become an excited state and vice versa. To verify it is so, first observe that the spin up and spin down number operators $\hat{N}^{\alpha} \equiv \int d^3\mathbf{r}(\psi^{\alpha}(\mathbf{r}))^{\dagger}\psi^{\alpha}(\mathbf{r})$, $\alpha = \uparrow, \downarrow$, commute with $\hat{H}_{(V;B)}$ and among themselves. Hence, the eigenstates of $\hat{H}_{(V;B)}$ must be (or, in case of degeneracy can be chosen to be) eigenstates of $\hat{N}^{\uparrow}, \hat{N}^{\downarrow}$. Shift now $V(\mathbf{r})$ by V_0 and $B(\mathbf{r})$ by B_0:

$$\hat{H}_{(V+V_0;B+B_0)} = \hat{H}_{(V;B)} + V_0 \hat{N} + \mu_0 B_0 (\hat{N}^{\uparrow} - \hat{N}^{\downarrow}). \tag{26}$$

If Ψ is an eigenstate of $\hat{H}_{(V;B)}$ with eigenvalue E,

$$\hat{H}_{(V;B)}|\Psi\rangle = E|\Psi\rangle, \tag{27}$$

we shall have:

$$\hat{H}_{(V+V_0;B+B_0)}|\Psi\rangle = (E + V_0 N + \mu_0 B_0 Z)|\Psi\rangle \tag{28}$$

where, $Z = N^{\uparrow} - N^{\downarrow}$. So, Ψ will be an eigenstate of $\hat{H}_{(V+V_0;B+B_0)}$ with eigenvalue $E + V_0 N + \mu_0 B_0 Z$. Note here that Ψ would not necessarily be an eigenstate of $\hat{H}_{(V+V_0;B+B_0)}$ if the magnetic field were not axial (because $\hat{N}^{\uparrow}, \hat{N}^{\downarrow}$ would not commute with the Hamiltonian).

If the constant shift B_0 is small enough that the lowest leves do not cross and the ground state remains a ground state, then the effect of B_0 will be trivial just like V_0. The value of the critical constant $\mu_0 B_{crit}$ necessary to induce a level crossing between the ground state and an excited state is the energy necessary to flip a spin and its physical meaning is similar to a chemical potential.

The non-uniqueness in B corresponds to and explains the non-uniqueness in the spin-potentials observed by Capelle and Vignale [6]. One may ask however, if this is all the freedom there is, or indeed if potentials in SDFT are truly not defined uniquely, in which case the theory cannot be formulated consistently. In other words, do essentially different spin potentials - i.e., potentials that differ by more than a spin-dependent constant - always lead to different spin densities? If we restrict ourselves to continuous potentials, we may give an affirmative answer and thereby complete the proof of the HK theorem for DSDFT.

Consider a continuous spin potential $(V^\uparrow, V^\downarrow)$ and $|\Psi\rangle$ an eigenstate of the spin Hamiltonian $\hat{H}_{(V^\uparrow, V^\downarrow)}$ in (6) for the case of diagonal spin-potentials. The Hamiltonian commutes with the spin up and spin down number operators and we take $|\Psi\rangle$ to describe a state with a definite number of spin up and spin down electrons $N^\uparrow, N^\downarrow = N - N^\uparrow$ with $N^\uparrow, N^\downarrow \neq 0$ (i.e. we also assume that the system is not completely polarized).

Theorem

It is impossible to find another continuous potential $(V'^\uparrow, V'^\downarrow)$ which has the same state $|\Psi\rangle$ as eigenstate and whose components differ from $(V^\uparrow, V^\downarrow)$ by more than a spin-dependent constant.

Proof

Assume the contrary, that a continuous potential $(V'^\uparrow, V'^\downarrow)$ with the same eigenstate exists. The two Schrödinger equations are,

$$(\hat{T} + \hat{W} + \sum_{\alpha=\uparrow,\downarrow} \hat{V}^\alpha)|\Psi\rangle = E|\Psi\rangle \qquad (29)$$

$$(\hat{T} + \hat{W} + \sum_{\alpha=\uparrow,\downarrow} \hat{V}'^\alpha)|\Psi\rangle = E'|\Psi\rangle, \qquad (30)$$

with $\hat{V}^\alpha = \int d^3\mathbf{r}\, V^\alpha(\mathbf{r})\hat{\rho}^\alpha(\mathbf{r})$, $\alpha = \uparrow,\downarrow$. Shift the potentials $(V^\uparrow, V^\downarrow)$ and $(V'^\uparrow, V'^\downarrow)$ by constants $-E/N$ and $-E'/N$ and subtract the two equations. One gets

$$\sum_{\alpha=\uparrow,\downarrow} \delta\hat{V}^\alpha |\Psi\rangle = 0 \qquad (31)$$

where $\delta\hat{V}^\alpha = \hat{V}'^\alpha - \hat{V}^\alpha$. We are assuming that $\delta V^\alpha(\mathbf{r})$ is not a constant.

Take the inner product of the lhs of the equation with $\langle 0|\hat{\psi}^\downarrow(\mathbf{r}_N) \ldots \hat{\psi}^\downarrow(\mathbf{r}_{N^\uparrow+1})\, \hat{\psi}^\uparrow(\mathbf{r}_{N^\uparrow}) \ldots \hat{\psi}^\uparrow(\mathbf{r}_1)$. Doing the commutators, we find

$$\left[\sum_{i=1}^{N^\uparrow} \delta V^\uparrow(\mathbf{r}_i) + \sum_{j=N^\uparrow+1}^{N} \delta V^\downarrow(\mathbf{r}_j)\right] \Psi(\mathbf{r}_1, \ldots \mathbf{r}_{N^\uparrow}; \mathbf{r}_{N^\uparrow+1}, \ldots, \mathbf{r}_N) = 0 \qquad (32)$$

where, $\Psi(\mathbf{r}_1 \ldots \mathbf{r}_{N^\uparrow}; \mathbf{r}_{N^\uparrow+1} \ldots \mathbf{r}_N) = (N^\uparrow! N^\downarrow!)^{-1/2} \langle 0|\, \hat{\psi}^\downarrow(\mathbf{r}_N) \ldots \hat{\psi}^\downarrow(\mathbf{r}_{N^\uparrow+1}) \hat{\psi}^\uparrow(\mathbf{r}_{N^\uparrow}) \ldots \hat{\psi}^\uparrow(\mathbf{r}_1) |\Psi\rangle$.

We will use the shorthand notation $\Psi(\mathbf{r}_1 \ldots \mathbf{r}_N)$ for $\Psi(\mathbf{r}_1 \ldots \mathbf{r}_{N\uparrow}; \mathbf{r}_{N\uparrow+1} \ldots \mathbf{r}_N)$ in the following. The normalized wavefunction $\Psi(\mathbf{r}_1 \ldots \mathbf{r}_N)$ is antisymmetric in $\mathbf{r}_i$, $1 \leq i \leq N^\uparrow$, and $\mathbf{r}_j$, $N^\uparrow + 1 \leq j \leq N$, but not with respect to an interchange of a 'spin up' and a 'spin down' coordinate $\mathbf{r}_i$ and $\mathbf{r}_j$. $\Psi(\mathbf{r}_1 \ldots \mathbf{r}_N)$ has the same spin up and spin down densities as the ground state $|\Psi\rangle$ of the full Hamiltonian and satisfies the Schrödinger equation:

$$[-\frac{\hbar^2}{2m}\sum_{k=1}^{N}\nabla_j^2 + \sum_{i=1}^{N^\uparrow}V^\uparrow(\mathbf{r}_i) + \sum_{j=N^\uparrow+1}^{N}V^\downarrow(\mathbf{r}_j)$$
$$+\frac{e^2}{2}\sum_{k\neq l=1}^{N}\frac{1}{r_{kl}}]\Psi(\mathbf{r}_1\ldots\mathbf{r}_N) = E\,\Psi(\mathbf{r}_1\ldots\mathbf{r}_N)$$

$\Psi(\mathbf{r}_1 \ldots \mathbf{r}_N)$ can be interpreted to describe a two-component system of spin 1/2 fermions.

Consider a region $\mathcal{R}$ in $\mathbf{R}^{3N}$, where $\mathbf{R}$ is the real axis, with the property that $\Psi(\mathbf{r}_1 \ldots \mathbf{r}_N)$ is nonzero for all $\{\mathbf{r}_1, \ldots \mathbf{r}_{N\uparrow}; \mathbf{r}_{N\uparrow+1}, \ldots, \mathbf{r}_N\}$ belonging in $\mathcal{R}$. Divide Eq. (32) through by $\Psi(\mathbf{r}_1 \ldots \mathbf{r}_N)$ to get

$$\sum_{i=1}^{N^\uparrow}\delta V^\uparrow(\mathbf{r}_i) + \sum_{j=N^\uparrow+1}^{N}\delta V^\downarrow(\mathbf{r}_j) = 0 \tag{33}$$

for all vectors $\mathbf{r}_i, \mathbf{r}_j$ which form points $\{\mathbf{r}_1, \ldots \mathbf{r}_{N\uparrow}; \mathbf{r}_{N\uparrow+1}, \ldots, \mathbf{r}_N\}$ of $\mathcal{R}$. Since the two terms correspond to different coordinates, we must have that each term separately equals a constant: $\sum_{i=1}^{N^\uparrow}\delta V^\uparrow(\mathbf{r}_i) = C$ and $\sum_{j=N^\uparrow+1}^{N}\delta V^\downarrow(\mathbf{r}_j) = -C$. By the same arguement, $\delta V^\uparrow(\mathbf{r}_i) = C/N^\uparrow$ and $\delta V^\downarrow(\mathbf{r}_j) = -C/N^\downarrow$ for any $\mathbf{r}_i, \mathbf{r}_j$.

If the region where Ψ is non-zero is bounded by a nodal surface, then $\mathbf{R}^{3N}$ will be separated in disconnected regions $\mathcal{R}$ and the corresponding constants C might be different in the different regions.

The vectors $\mathbf{r}_i$ and $\mathbf{r}_j$ which compose the $3N$–dimensional vectors of $\mathcal{R}$ will form the regions $\mathcal{R}^\uparrow$ and $\mathcal{R}^\downarrow$ in $\mathbf{R}^3$: $\mathbf{r}_i \in \mathcal{R}^\uparrow$ and $\mathbf{r}_j \in \mathcal{R}^\downarrow$. Vectors $\mathbf{r}_i$ and $\mathbf{r}_j$ composing $3N$–dimensional vectors in a different region $\mathcal{R}'$ will form different in general regions $\mathcal{R}^\uparrow{}'$ and $\mathcal{R}^\downarrow{}'$ of $\mathbf{R}^3$. If either $\mathcal{R}^\uparrow$ and $\mathcal{R}^\uparrow{}'$, or, $\mathcal{R}^\downarrow$ and $\mathcal{R}^\downarrow{}'$ have common points then the constants C and C' will be the same. Assume that there are two regions $\mathcal{R}^\uparrow$ and $\mathcal{R}^\uparrow{}'$ with no common points. We must be able to find two such regions which are neighbouring, because all $\mathbf{R}^3$ must be covered. Then for points $\mathbf{r}, \mathbf{r}'$ belonging in different $\mathcal{R}^\uparrow$ and $\mathcal{R}^\uparrow{}'$ and arbitrarily close, $\delta V^\uparrow(\mathbf{r}) = C/N^\uparrow$ and $\delta V^\uparrow(\mathbf{r}') = C'/N^\uparrow$, which implies that the potential $V'^\uparrow$ must be discontinuous, since it must equal $V^\uparrow(\mathbf{r})+C/N^\uparrow$ and $V^\uparrow(\mathbf{r}')+C'/N^\uparrow$ for points $\mathbf{r}$, $\mathbf{r}'$ which lie arbitrarily close, and the proof is complete.

Before we proceed, it is worthwhile to note the similarity between the formulation of SDFT and the formulation of two-component DFT [12]. In addition, the proof of the theorem can be modified easily to establish the 1-1 correspondence between potentials and densities in two-component DFT.

Having established that different spin-potentials always correspond to different densities and inversely, one may try to incorporate in the constrained search formulation of DSDFT that spin-potentials should be considered different, if they differ by more than a spin-constant. Remember that the potential arises as the partial functional derivative of a spin-density functional. Eq. (25) on the contrary, suggests that the potential can be determined within an overall constant. This is the result of introducing one constraint (23) and obviously, if we want two constants to appear, we should use two constraints, namely, that the components of the spin-density integrate separately to fixed number of electrons. This is consistent with the observation that the number of up and down electrons are good quantum numbers and the remaining thing to worry about is that the true number of spin up and down electrons for the ground state should be determined variationally. This can be accomplished by constraining further the constrained search scheme. Since the number of up and down electrons are good quantum numbers, in the search for the ground state, one needs to consider only the states in the N-particle Hilbert space with a definite number of up and down electrons along the direction of the magnetic field. Their space can be divided to $N+1$ disjoint subspaces with number of up electrons running from 0 to N. We then search separately for the state with the lowest energy in each of these subspaces. The state with the lowest energy of all, will be the ground state we seek.

We can formally write these minimal energies in the disjoint spaces as:

$$E_V[N^\uparrow, N^\downarrow] = \min_{\Psi \to N^\uparrow, N^\downarrow} \langle \Psi | H_V | \Psi \rangle \qquad (34)$$

We can then employ a constrained search scheme for each of these minimal energies $E_V[N^\uparrow, N^\downarrow]$:

$$E_V[\rho^\uparrow, \rho^\downarrow] = F[\rho^\uparrow, \rho^\downarrow] + \sum_{\alpha=\uparrow,\downarrow} \int d^3\mathbf{r}\, V^\alpha(\mathbf{r})\rho^\alpha(\mathbf{r}) \qquad (35)$$

$$F[\rho^\uparrow, \rho^\downarrow] = \min_{\Psi \to \rho^\uparrow, \rho^\downarrow} \langle \Psi | \hat{T} + \hat{W} | \Psi \rangle \qquad (36)$$

$$E_V[N^\uparrow, N^\downarrow] = \min_{(\rho^\uparrow, \rho^\downarrow) \to (N^\uparrow, N^\downarrow)} E_V[\rho^\uparrow, \rho^\downarrow] \qquad (37)$$

The true ground state energy and the correct number of up and down electrons can be found by picking the smallest of the $N+1$ numbers:

$$E_\mathcal{V} = \min_{N^\uparrow=0,\ldots,N} E_\mathcal{V}[N^\uparrow, N - N^\uparrow] \qquad (38)$$

Note that the definition of the functionals $E_\mathcal{V}[\rho^\uparrow, \rho^\downarrow]$, $F[\rho^\uparrow, \rho^\downarrow]$ is not different from the traditional one in (19,20). What this reformulation suggests is that the arguments $\rho^\uparrow$, $\rho^\downarrow$ of the functionals are constrained to integrate to given numbers of particles and hence, upon taking the functional derivative, the singly constrained derivatives (i.e. constraining only the sum $\rho^\uparrow + \rho^\downarrow$ to integrate to N) will not exist in many cases, while, it is reasonable to assume that the doubly constrained functional derivative will exist. An assumption which is supported by the HK theorem for DSDFT and which is therefore completely analogous to the spin-independent case.

The constrained search formulation is considered to offer an equivalent formulation of DFT, independent of the proof of the HK theorem. Although this may be true, an open question always remains, which is the assumed existence of the functional derivative of the universal functional F with respect to the density. We realise that this assumption must be consistent with a HK theorem that ensures the invertibility of the mapping between potentials and densities.

References

[1] W. Kohn, L.J. Sham, *Phys. Rev.* 140, A1133 (1965).
[2] P. Hohenberg, W. Kohn, *Phys. Rev.* **136B**, 864 (1964).
[3] M. Levy, *Proc. Natl. Acad. Sci. USA*, **76**, 6062 (1979); *Phys. Rev. A* **26**, 1200 (1982).
[4] E.H. Lieb, in *Physics as a Natural Philosophy*, ed. A. Shimony, H. Feshbach, MIT Press, Cambridge, p.111 (1982); in *Int. J. Quant. Chem.* **24**, 243 (1983); in *Density Functional Methods in Physics*, NATO ASI Series B123, ed. R.M. Dreizler, J. da Providencia, Plenum, New York, p31 (1985).
[5] J.P. Perdew and A. Zunger, Phys. Rev. B **23**, 5048 (1981).
[6] K. Capelle and G. Vignale, *Phys. Rev. Lett.*,**86**, 5546 (2001); *Phys. Rev. B*, **65**, 113106 (2002)
[7] U. von Barth and L. Hedin, *J. Phys. C* **5**, 1629 (1972).
[8] M.M. Pant and A.K. Rajagopal, *Sol. State Comm.* **10**, 1157 (1972).
[9] R.M. Dreizler, E.K.U. Gross, *Density Functional Theory*, Springer Verlag, Berlin, (1990).
[10] H. Eschrig and W.E. Pickett, *Solid State Commun.* **118**, 123 (2001)
[11] N.I. Gidopoulos, to appear
[12] N. Gidopoulos, *Phys. Rev. B* **57**, 2146 (1998)

III

THE FORUM

THE FORUM - QUESTIONS

Participants (and others who had expressed an interest) were invited to submit questions for discussion at the Forum prior to the meeting. The questions received were distributed by e-mail to all participants (and to others who had expressed an interest) two weeks before the meeting. The questions received were as follows:-

1. Everything is a unique functional of the density – but what density, the actual density or the model density ?
   And how are the two related ?
   R. McWeeny

2. What is the best approximate correlation potential with a Krieger-Lee-Iafrate (KLI) exchange potential ?
   A Nagy

3. I've always wanted to know how one might use a density functional approach to do a molecular calculation while allowing the nuclei to move freely. Specifically, suppose that one repeated Kolos and Wolniewicz's calculation on H_2 but used a density functional theory approach. How precisely would one do it ?
   B T Sutcliffe

4. Can one prove the existence of the demonstrated local relation between the electron density and the electronic kinetic and potential energy densities, all quantities being determined by the one-electron density matrix ?
   R F W Bader

5. Can one develop a density functional theory for phenomena that do not conserve particle number, such as photoemission ?
   R van Leeuwen

6 In a real atom with more than one electron, are there s, p, d, ... orbitals ?
How are they defined ?
's electrons' ?
In a Hartree-Fock wave function ?
We can put an electron into an atom, and we can get one out. Are there any inside?
A J Coleman

7 Is it possible to make a whole molecular orbitals set from the total electron density?
J J Lee

8 What is the best way to present density functional theory to an undergraduate audience?
P Madden

9 In a paper published in 1998, E R Davidson asked the question *"How robust is Present-Day DFT ?"* He writes

> "present-day ... functionals give useful results for neutral closed-shell molecules of low atomic number. For high Z, high charge, or systems with large configuration mixing, present functional are less useful."

Has the situation changed?
S Wilson

10 Starting from a single determinant model function, how can one get the charge density, spin density, and current density for the actual system?
R. McWeeny

THE FORUM - DISCUSSION

The Forum was chaired by Professor B.T. Sutcliffe. The following summary of the discussion was compiled with the help of the participants after the meeting.

Question 1

Everything is a unique functional of the density – but what density, the actual density or the model density ?
And how are the two related ?

R. McWeeny

Discussion 1

Ludena:

Given any model density $\rho_\Psi(\mathbf{r}_1)$ obtained from the model wavefunction $\Psi(\mathbf{r}_1, ..., \mathbf{r}_N)$ by $\rho_\Psi(\mathbf{r}) = N \int d\mathbf{r}_2 ... \int d\mathbf{r}_N \Psi^* \Psi$ and the exact density $\rho_0^{exact}(\mathbf{r})$ corresponding to the exact ground-state wavefunction $\Psi_0^{exact}(\mathbf{r}_1, ..., \mathbf{r}_N)$, then through the application of local-scaling transformations connecting the densities ρ_Ψ and ρ_0^{exact} it is possible to generate a locally-scaled model wavefunction $\Psi([\rho_0^{exact}]; \mathbf{r}_1, ..., \mathbf{r}_N)$ that associates with the exact density ρ_0^{exact}. Conversely, the exact wavefunction can be locally scaled so as to produce transformed wavefunctions that associate with any admissible density. The problem in DFT, in our opinion, is not related to the type of density one must use, but rather to how one represents the energy as a functional of the one-particle density, so that the extremum of this functional satisfies the variational principle.

Levy:

Please define the model density.

McWeeny:

The model density is the density associated with the non-interacting system represented by the Slater determinant.

Levy:

We know from Harriman's work (Phys. Rev. A, **24**, 680 (1981)) that the two densities, the density of the model system and the actual density can be the same. This does not imply that the density matrices coincide.

Additional comments
Goerling:

It is important to distinguish between the first order density matrix and the electron density. The many-particle Kohn-Sham wave function, in the absence of degeneracies is a single Slater determinant. Its first order density matrix (the first order Kohn-Sham model density matrix) thus has eigenvalues which are either zero or one. Therefore it can not be identical to the first order density matrix of the real interacting system (the actual first order density matrix) which, in general, has eigenvalues between zero and one. The Kohn-Sham electron density, i.e., the trace of the first order Kohn-Sham density matrix, on the other hand, by construction, equals the actual electron density of the real interacting system. Remember that different first order density matrices can have the same trace. Of course, the kinetic energy of the Kohn-Sham system differs from that of the real system because the first order density matrices of the two systems are different. This difference in the kinetic energy is accounted for in the Kohn-Sham correlation energy. Note that the definition of the later differs from the Hartree-Fock based definition of the correlation energy.

Question 2

What is the best approximate correlation potential with a Krieger-Lee-Iafrate (KLI) exchange potential ?

A Nagy

Discussion 2
Ludena:

The correlation energy is made up of the non-dynamical and thedynamical components. A rigorous definition of these components hasbeen given by Cioslowski (J. Cioslowski, Phys. Rev. A **43**, 1223(1991)). An analysis of these components for two and four electronatoms has been given by Valderrama et al. (E. Valderrama,E.V. Ludeña and J. Hinze, J. Chem. Phys. **106**, 9227 (1997);**110**, 2343 (1999)). By and large, correlation functionals only incorporate the dynamical part. The non-dynamical contribution hasbeen introduced in a more or less unintended way through the exchange term. When the exact exchange (or highly approximate exchange as in the KLI method) is used, then the spurious correction describing the non-dynamical correlation is eliminated. The solution would be to purposely design non-dynamical correlation functionals following, perhaps, the approach presented by Valderrama et al.

Theophilou:

In a paper with P. Papaconstantinou (to be submitted) we found that for atoms the best correlation energy expression is the Cole Salvetti one (C.R. Colle and D. Salvetti, Theor. Chim. Acta **37**, 329 (1975), which gives 0.00 to 0.05 per cent energy deviation. In fact our application was not exactly on the Krieger-Lee-Iafrate approach, but a parametrized form of it. However for molecules, we do not know any sufficiently good correction.

Question 3

I've always wanted to know how one might use a density functional approach to do a molecular calculation while allowing the nuclei to move freely. Specifically, suppose that one repeated Kolos and Wolniewicz's calculation on H_2 but used a density functional theory approach. How precisely would one do it ?

B T Sutcliffe

Discussion 3

Sutcliffe:

There have been two contributions at this meeting dealing with this problem.

Gidopoulos:

My work with Dr. Kohanoff does not constitute a density functional treatment of the nuclear degrees of freedom. It is a wave function approach which by introducing a further approximation, enables one to apply the Born-Oppenheimer scheme to more complex systems than was previously possible.

The theory presented by Prof. Gross on the other hand, provides a true density functional treatment of the nuclear degrees of freedom. The test case was H_2^+ (see T. Kreibich and E. K. U. Gross, *Multicomponent Density-Functional Theory for Electrons and Nuclei*, Phys. Rev. Lett. **86**, 2984 (2001)). A one-electron system was used to remove all electron-electron correlation effects allowing to focus only on the electron-nuclear correlation. The calculation is directly comparable with the work of Kolos and Wolniewicz.

Question 4

Can one prove the existence of the demonstrated local relation between the electron density and the electronic kinetic and potential energy densities, all quantities being determined by the one-electron density matrix ?

R F W Bader

Discussion 4

Theophilou:

One can get local energy density functionals like the kinetic energy one, by functional differentiation of the corresponding global functionals, but I doubt about the existence of these functional derivatives. In fact there is an interesting discussion nowadays on this matter. See e.g. R.K. Nesbet, Phys. Rev. A **65**, 010502(R) (2001), *Ibid*: A **58**, R12 (1998); A. Holas and N.H. March, Phys. Rev. A **64,** 016501 (2001), *Ibid: A* **66**, 066501 (2002).

Ludena:

There is a very interesting decomposition of the one-particle density which can be attained via the first-order Density Equation of Nakatsuji (see, for example, H. Nakatsuji, in *Many-Electron Densities and Reduced Density Matrices*, edited by J. Cioslowski, (Kluwer Academic/Plenum Publishers, New York, 2000) p. 85). For Hooke's atom, this connection can be explicitly studied by means of analytic expressions (see A. Artemyev, E.V. Ludeña and V. Karasiev, Theochem **580**, 47 (2002)). Using this same decomposition, it is probably feasible to relate aspects of the topology of ρ to those of the external potential, and the kinetic energy density.

Additional comment

Gidopoulos:

The dependence of the internal (kinetic plus Coulomb repulsion) energy functional, $F[\rho] \equiv \min_{\Psi \to \rho} \langle \Psi | T + W | \Psi \rangle$ as defined by Levy (Proc. Natl. Acad. Sci. U.S.A., **76**, 6062 (1979)) and Lieb (Int. J. Quantum Chem., **24**, 243 (1983)) on the density, is certainly not local. The existence of the functional derivative $\frac{\delta F[\rho]}{\delta \rho(\mathbf{r})}$ is an open issue in the formulation of DFT (the only one, I would claim) and must be assumed. If $\frac{\delta F[\rho]}{\delta \rho(\mathbf{r})}$ does exist however, it will necessarily be a function of $\mathbf{r}$ only and will play the role of a local effective potential. Again, this potential, viewed as a density functional, does not depend on the density locally.

Question 5

Can one develop a density functional theory for phenomena that do not conserve particle number, such as photoemission ?

R van Leeuwen

Discussion 5

Theophilou:

One may formulate an ensemble theory where the elements of the ensemble have differing particle numbers.

Gross:

The natural treatment of such processes is with the one-body Green's function. Its poles are the "non-neutral excitation energies" $E(N) - E(N-1)$ and $E(N) - E(N+1)$, i.e. exactly the quantities measured in photoelectron spectroscopy. DFT does not treat these quantities in a natural way. Of course, one might use TDDFT. For a finite system, the wave function is, at any finite time, normalized to 1, and the same holds true for the TDKS orbitals. Hence, photoelectron spectra are within the realm of TDDFT, at least in principle. In practice, however, it is very hard to find good functionals of the time-dependent density describing the photoelectron spectrum. Some first attempts to describe at least the ionisation yields were made in M. Petersilka and E.K.U. Gross, Laser Physics **9**, 105 (1999).

Question 6

In a real atom with more than one electron, are there s, p, d, ... orbitals ?
How are they defined ?
's electrons' ?
In a Hartree-Fock wave function ?
We can put an electron into an atom, and we can get one out. Are there any inside?

A J Coleman

Discussion 6

McWeeny:

We can define the orbitals rigorously by looking at the eigenfunctions of the single particle density matrix. This defines the natural orbitals.

Gidopoulos:

Do the natural orbitals preserve the s, p, d, ... symmetry ?

McWeeny:

Yes they do.

Theophilou:

For a nondegenerate state one can always reproduce the density by using a Slater derivative, consisting of one particle eigenfunctions of angular momentum. However, for the case of degenerate states the density does not have spherical symmetry. Fertig and Kohn have given an example of an excited state density, where the exact many particle wave function is an eigenstate of the total angular momentum, the density however cannot be produced by a single determinant having this symmetry. Then the spin orbitals are linear combinations of different angular momentum orbitals see H.A. Fertig and W. Kohn, phys. Rev. A **62**, 052511 (2000).

Goerling:

If chemists think of electronic structures they think in terms of orbitals, e.g., if one characterizes excitations or changes of the electronic structures during reactions. Thus orbitals provide an enormous insight, chemical intuition to some extent is built on orbitals. Also characterizations of electronic sconfigurations or the concept of statical and dynamical correlation depend on orbitals. The question, of course, is: Which orbitals? Certainly not Hartree-Fock orbitals because usually there are no or just very few bound unoccupied Hartree-Fock orbitals. Kohn-Sham orbitals, on the other hand, seem to reflect chemical intuition and seem to represent a sound basis for the characterization of electronic structures provided they are obtained in a Kohn-Sham approach free of Coulomb self-interaction, e.g., by (effective) exact-exchange Kohn-Sham procedures.

Question 7

Is it possible to make a whole molecular orbitals set from the total electron density?

J J Lee

Discussion 7

Goerling:

There exist various procedures to obtain the effective Kohn-Sham potential and thus the Kohn-Sham orbitals from a given electron density [among others: Q. Zhao and R.G. Parr, Phys. Rev. A 46, 2337 (1992); Q. Zhao, R.C. Morrison, and R.G. Parr, Phys. Rev. A 50, 2138 (1994);

R. v. Leeuwen and E. J. Barends, Phys. Rev. A 49, 2421 (1994); P.R.T. Schipper, O. V. Gritsenko, and E. J. Baerends, Theor. Chem. Acc. 98, 16 1997]. A difficulty occuring in the construction of the effective Kohn-Sham potential from a given electron density is the numerical instability of the problem, i.e., small errors or inaccuracies in the electron density can lead to huge errors in the effective potential which may strongly affect the unoccupied orbitals.

Question 8

What is the best way to present density functional theory to an undergraduate audience?

P Madden

Editorial Note: This question was not discussed at the meeting because of lack of time.

Question 9

In a paper published in 1998, E R Davidson asked the question *"How robust is Present-Day DFT ?"* He writes

"present-day ... functionals give useful results for neutral closed-shell molecules of low atomic number. For high Z, high charge, or systems with large configuration mixing, present functional are less useful."

Has the situation changed?

S Wilson

Discussion 9

Theophilou:

As I stated earlier I have some doubts about the existence of functional derivatives. However, I could reproduce Kohn's density functional theory by defining functionals not in the space of densities but in the whole Hilbert space. In this formulation no functional derivatives are used. See A.K. Theophilou, Int. Jour. of Quantum Chem. **69**, 461 (1998).

Question 10

Starting from a single determinant model function, how can one get the charge density, spin density, and current density for the actual system?

R. McWeeny

Discussion 10

Editorial Note: This question was not discussed at the meeting because of lack of time.

Comment

Gidopoulos:

Physical quantities in DFT are determined only as functionals of the density: $Q[\rho] \equiv \langle \Psi[\rho] | \hat{Q} | \Psi[\rho] \rangle$, where $\hat{Q}$ is the relevant operator and $\Psi[\rho]$ the minimizing state of

$$F[\rho] = \min_{\Psi \to \rho} \langle \Psi | T + W | \Psi \rangle$$

In the absence of appropriate approximate functionals for $Q[\rho]$, the Kohn-Sham (KS) Slater determinant $\Phi[\rho]$, (i.e., the minimizing Slater determinant of $T_s[\rho] = \min_{\Phi \to \rho} \langle \Phi | T | \Phi \rangle$), is often used, as an approximation to the interacting state $\Psi[\rho]$ in order to calculate the expectation value of $\hat{Q}$. It should be remembered however, that even in the limit when the interaction is almost switched off, $\Phi[\rho]$ can be a very poor approximation to the physical wavefunction, since it may not have the correct symmetry. Prof. Mc Weeny's contribution to this volume demonstates this point very clearly. Conceptually this is not worrying as far as the foundations of DFT are concerned, but it does imply that approximating the functional $Q[\rho]$ can be very hard indeed (usually a very complicated and non-local functional). The DFT community has found an elegant way around the problem of approximating $\hat{Q}$ in some special cases. When the quantity of interest Q is the spin-density for example, one defines an additional potential conjugate to Q, here the spin-density, and employs Spin-DFT (please see my contribution for a review of SDFT and further references). In the corresponding KS scheme of SDFT, the interacting system (where a magnetic field may be completely absent) is mapped to a non-interacting system, in such a way that the spin-densities of the two systems are the same. In order to satisfy this requirement, the effective potential of the non-interacting system must be spin-dependent. Similarly, when the quantity of interest is the current density, one invokes a mapping of the interacting system to a non-interacting one with the same current density. The reader is referred to R. van Leeuwen's contribution in this volume for a review of Current-DFT.

A GLOSSARY

Every field of scientific research requires the appropriate language and this inevitably introduces abbreviations for often used terms and phrases. Here we list alphabetically the abbreviations employed in this volume.

1DHM one-dimensional Hubbard Model

1-DM one-electron Density Matrix

2-DM two-electron Density Matrix

BA Bethe Ansatz

BA-LDA Bethe Ansatz-Local Density Approximation

CI Configuration Interaction

CSE Contracted Schrödinger Equation

DFT Density Functional Theory

DM Density Matrix

DMRG Denisty Matrix Renormalization Group

DMT Density Matrix Theory

DSDFT Diagonal Spin Density Functional Theory

GGA Generalized Gradient Correction

HK Hohenberg-Kohn

IPM Independent Particle Model

KC Kummer Cone

KS Kohn-Sham

KV Kummer Variety

LDA Local Density Approximation

LDA+U Local Density Approximation with Hubbard U

LSDA Local Spin Density Approximation

QMC Quantum Monte Carlo

SDFT Spin Density Functional Theory

SDW Spin-Density Wave

SLD Slater Determinant

SSMTG Solid State and Molecular Theory Group

TDCDFT Time-Dependent Current-Density Functional Theory

TDDFT Time-Dependent Density Functional Theory

TDOPM Time-Dependent Optimized Potential Method

WDA Weighted Density Approximation

A SELECTED BIBLIOGRAPHY

We include here a short selected bibliography of texts covering the theoretical study of electron density, density matrices and density functional theory.

Cioslowski, J., editor, (2000) *Many-Electron Densities and Reduced Density Matrices* Kluwer Academic Press, New York.

Coleman, A.J., and Yukalov, V.I., (2000) *Reduced Density Matrices: Coulson's Challenge* Lecture Notes in Chemistry Springer-Verlag, New York.

Davidson, E.R., (1976) *Reduced Density Matrices in Quantum Chemistry*, Academic Press, New York.

Dobson, J.F., editor, (1998) *Electronic Density Functional Theory: Recent Progress and New Directions*, Kluwer Academic Publisher, Dordrecht.

Gross, E.K.U., and Dreizler, R.M., editors, (1995) *Density Functional Theory*, NATO ASI Series B: Physics Vol. 337, Plenum Press, New York.

Koch, W., and Holthausen, M.C., (2000) *A Chemist's Guide to Density Functional Theory*, Wiley-VCH.

Kryachko, E.S., and Ludeña, E.V., (1990) *Energy Density Functional Theory of Many-Electron Systems*, Kluwer Academic Publisher, Dordrecht.

McWeeny, R., (1989) *Methods of Molecular Quantum Mechanics*, Academic Press, London.

Parr, R.G., and Yang, W., (1989) *Density-Functional Theory of Atoms and Molecules*, International Series of Monographs on Chemistry 16, Clarendon Press, Oxford.

Petkov, I. Zh., and Stoitsov, M.V., (1991) *Nuclear Density Functional Theory*, Oxford Studies in Nuclear Physics, Clarendon Press, Oxford.

Index

1-electron density matrix 97
1-matrix 129
2-electron density matrix 97

A
action functional 43
average field 185

B
Bethe-Ansatz-Local-Density Approximation (BA-LDA) 145

C
conditions on 1-matrix 129
conditions on reduced single-particle operator 129
constrained search formulation 195
constructive method 129
cosine-wave electron gas 172

D
density functional theory 129
density functional theory 5
density functional theory for the Hubbard model 145
density functional theory, functional N-representability in 129
density functional theory, new formulation of 115
density functional theory, pair 79
density functional theory, unsolved problems in 97
density matrix 79
density matrix theory 129
density matrix theory 4
density matrix theory 97
density matrix theory, functional N-representability in 129
density matrix theory, unsolved problems in 97
diagonal spin density functional theory 195
diagonal spin-potential 195
dissociation 69

E
electron density 3
electron gas, cosine wave 172
electron gas, exchange-only kernel for the 58
ensemble N-representable 129
exchange-correlation contribution to vector potential 43
exchange-correlation energy of the Hubbard model 145, 149
exchange-only kernel for the electron gas 58
external potential, Kohn-Sham potential as a mapping of 126

F
Fermi-liquid 145
functional derivative 115
functional group 185
functional N-representability 129

G
generalised gradient approximation (GGA) 169
group theory 108

H

Hartree-Fock equations 185
Hohenberg-Kohn theorem 195
Hooke's atom 129, 136
Hubbard model, density functional theory for 145
Hubbard model, exchange correlation energy of 145, 149

I

impurity models 156
inhomogeneous electron gas systems 169
Integral equation of the xc-kernel 55
invertible mapping 195
IPM approximation 97

K

Keldysh action 45
Keldysh formalism 43
Keldysh time contour 43
Kohn-Sham equations 49
Kohn-Sham equations of time-dependent current-density functional theory 43
Kohn-Sham potential 126
Kummer variety for N-particles 89

L

large amplitude perturbations 177
laser pulses 69
linear response, Kohn-Sham equations and 49
local density approximation 169
local-density approximation 145
local-scaling transformations 129, 137
long-range 185
Luttinger liquids 145, 154

M

molecules in strong laser pulses 69 insulator 145, 162

N

N-representability 97, 129
N-representability conditions 89

new formulation of density functional theory 115
nonlocal density functional description 169
nonlocal exchange correlation functional 169
nuclear degrees of freedom 69

O

one-dimensional model of H2+ 69
one-matrix 185

P

pair correlation functions 169, 177
pair density 79
pair density functional theory 79
physically admissible solution 97

R

reduced matrix theory 129
response theory 43

S

self-consistent field calculation 185
short-range 185
space and spin variables 97
spin density functional theory 195
spin-density 195
spin-density waves 145, 160
spin-potential 195
stationary-action principle 69
strong laser pulses 69
strongly inhomogeneous density 169
symmetry requirements 115

T

TDOPM equations 52
theory of atoms in molecules 185
time-dependent current density functional theory 43
time-dependent density functional theory 69
time-dependent optimized potential method 52
transferability of electron density 185
transferable properties 185
two-particle equation 79

INDEX

U
unsolved problems in density matrix theory and density functional theory 97

V
variational approach 69
variational collapse 97
variational Monte-Carlo (VMC) simulations 169

virial field 185
virial of Ehrenfest force 185

W
weighted density approximation (WDA) 169

X
xc-kernel, integral equation for 55

Progress in Theoretical Chemistry and Physics

1. S. Durand-Vidal, J.-P. Simonin and P. Turq: *Electrolytes at Interfaces.* 2000
   ISBN 0-7923-5922-4
2. A. Hernandez-Laguna, J. Maruani, R. McWeeny and S. Wilson (eds.): *Quantum Systems in Chemistry and Physics.* Volume 1: Basic Problems and Model Systems, Granada, Spain, 1997. 2000   ISBN 0-7923-5969-0; Set 0-7923-5971-2
3. A. Hernandez-Laguna, J. Maruani, R. McWeeny and S. Wilson (eds.): *Quantum Systems in Chemistry and Physics.* Volume 2: Advanced Problems and Complex Systems, Granada, Spain, 1998. 2000   ISBN 0-7923-5970-4; Set 0-7923-5971-2
4. J.S. Avery: *Hyperspherical Harmonics and Generalized Sturmians.* 1999
   ISBN 0-7923-6087-7
5. S.D. Schwartz (ed.): *Theoretical Methods in Condensed Phase Chemistry.* 2000
   ISBN 0-7923-6687-5
6. J. Maruani, C. Minot, R. McWeeny, Y.G. Smeyers and S. Wilson (eds.): *New Trends in Quantum Systems in Chemistry and Physics.* Volume 1: Basic Problems and Model Systems. 2001   ISBN 0-7923-6708-1; Set: 0-7923-6710-3
7. J. Maruani, C. Minot, R. McWeeny, Y.G. Smeyers and S. Wilson (eds.): *New Trends in Quantum Systems in Chemistry and Physics.* Volume 2: Advanced Problems and Complex Systems. 2001   ISBN 0-7923-6709-X; Set: 0-7923-6710-3
8. M.A. Chaer Nascimento: *Theoretical Aspects of Heterogeneous Catalysis.* 2001
   ISBN 1-4020-0127-4
9. W. Schweizer: *Numerical Quantum Dynamics.* 2001   ISBN 1-4020-0215-7
10. A. Lund and M. Shiotani (eds.): *ERP of Free Radicals in Solids.* Trends in Methods and Applications. 2003   ISBN 1-4020-1249-7
11. U. Kaldor and S. Wilson (eds.): *Theoretical Chemistry and Physics of Heavy and Superheavy Elements.* 2003   ISBN 1-4020-1371-X
12. J. Maruani, R. Lefebvre and E. Brändas (eds.): *Advanced Topics in Theoretical Chemical Physics.* 2003   ISBN 1-4020-1564-X
13. J. Rychlewski (ed.): *Explicitly Correlated Wave Functions in Chemistry and Physics.* Theory and Applications. 2003   ISBN 1-4020-1674-3
14. N.I. Gidopoulos and S. Wilson (eds.): *The Fundamentals of Electron Density, Density Matrix and Density Functional Theory in Atoms, Molecules and the Solid State.* 2003
    ISBN 1-4020-1793-6

KLUWER ACADEMIC PUBLISHERS – DORDRECHT / LONDON / BOSTON